▶ **The Role of Creative Ignorance**

DOI: 10.1057/9781137492470.0001

Other Palgrave Pivot titles

James Carson: The Columbian Covenant: Race and the Writing of American History

Tomasz Kamusella: Creating Languages in Central Europe during the Last Millennium

Imad A. Moosa and Kelly Burns: Demystifying the Meese–Rogoff Puzzle

Kazuhiko Togo and GVC Naidu (editors): Building Confidence in East Asia: Maritime Conflicts, Interdependence and Asian Identity Thinking

Aylish Wood: Software, Animation and the Moving Image: What's in the Box?

Mo Jongryn (editor): MIKTA, Middle Powers, and New Dynamics of Global Governance: The G20's Evolving Agenda

Holly Jarman: The Politics of Trade and Tobacco Control

Cruz Medina: Reclaiming Poch@ Pop: Examining the Rhetoric of Cultural Deficiency

David McCann: From Protest to Pragmatism: The Unionist Government and North-South Relations from 1959–72

Thijl Sunier and Nico Landman: Transnational Turkish Islam: Shifting Geographies of Religious Activism and Community Building in Turkey and Europe

Daria J. Kuss and Mark D. Griffiths: Internet Addiction in Psychotherapy

Elisa Giacosa: Innovation in Luxury Fashion Family Business: Processes and Products Innovation as a Means of Growth

Domagoj Hruška: Radical Decision Making: Leading Strategic Change in Complex Organizations

Bjørn Møller: Refugees, Prisoners and Camps: A Functional Analysis of the Phenomenon of Encampment

David Ralph: Work, Family and Commuting in Europe: The Lives of Euro-commuters

Emily F. Henderson: Gender Pedagogy: Teaching, Learning and Tracing Gender in Higher Education

Mihail Evans: The Singular Politics of Derrida and Baudrillard

Bryan Fanning and Andreas Hess: Sociology in Ireland: A Short History

Tom Watson (editor): Latin American and Caribbean Perspectives on the Development of Public Relations: Other Voices

Anshu Saxena Arora and Sabine Bacouël-Jentjens (editors): Advertising Confluence: Transitioning the World of Marketing Communications into Social Movements

DOI: 10.1057/9781137492470.0001

palgrave▸pivot

The Role of Creative Ignorance: Portraits of Path Finders and Path Creators

Piero Formica

Innovation Value Institute, Maynooth University, Ireland

palgrave
macmillan

DOI: 10.1057/9781137492470.0001

First published in 2015
by PALGRAVE MACMILLAN®
in the United States—a division of St. Martin's Press LLC,
175 Fifth Avenue, New York, NY 10010.

Where this book is distributed in the UK, Europe and the rest of the world,
this is by Palgrave Macmillan, a division of Macmillan Publishers Limited,
registered in England, company number 785998, of Houndmills,
Basingstoke, Hampshire RG21 6XS.

Palgrave Macmillan is the global academic imprint of the above companies
and has companies and representatives throughout the world.

Palgrave® and Macmillan® are registered trademarks in the United States,
the United Kingdom, Europe and other countries.

ISBN: 978-1-137-49248-7 EPUB
ISBN: 978-1-137-49247-0 PDF
ISBN: 978-1-137-48962-3 Hardback

Library of Congress Cataloging-in-Publication Data is available from
the Library of Congress.

A catalogue record of the book is available from the British Library.

First edition: 2015

www.palgrave.com/pivot

DOI: 10.1057/9781137492470

To the thread of Ariadne to which all of us are tied.

To Martin Curley, the ideal companion on journeys in the unexplored.

DOI: 10.1057/9781137492470.0001

Contents

DOI: 10.1057/9781137492470.0001

DOI: 10.1057/9781137492470.0001

Foreword

The poet Antonio Machado wrote, 'Traveller, there is no path. The path is made by walking. By walking you make a path' (Machado, 2003). In the history of our human society path finders and path creators have played a huge and often unheralded role in advancing civilization and progress. Whether it was Christopher Columbus' discovery of America, epic arctic journeys of discovery to Antarctica by Ernest Henry Shackleton and others, or Thomas Edison's invention of the light bulb, these path explorers have expended great energy, taken great risks, experienced pain and ecstasy, put their reputations and sometimes their lives on the line. Almost always they were driven by a compelling vision and intuitive hunches and we have all benefitted from the paths they found and created. However, each new discovery or path found relies on previous paths found and created—for example, others such as Joseph Swan, Edward Weston and Hiram Maxim helped pave the way for the light bulb invention. As Sir Isaac Newton said, 'if I have seen further it is by standing on the shoulders of giants'.

We all owe these extraordinary path finders and path creators an enormous debt of gratitude. How many times have we driven on a road or motorway without thinking of the labor and toil of the workers who built the road and the planners who envisaged the route? It is the normal path of innovation and creation that new innovations are built on the creation and achievements of others.

What drives these path finders and path creators? Don Miquel Ruiz's statement that 'All of humanity is searching

DOI: 10.1057/9781137492470.0002

for truth, justice and beauty' is one potential motivation. While 'necessity is known as the mother of invention', we know that 'curiosity' is a very important driver for path creators. However the saying 'curiosity killed the cat' reminds us that path finding and creation are not without danger. Some, like the famous polar explorer Sir Robert Scott, paid the price with their lives.

We also know that path finders and path creators are inherently optimistic. While pessimists see difficulty in every opportunity, optimists see opportunity in every challenge. The founder of Intel, Robert Noyce, often stated that 'optimism is an essential ingredient for Innovation'. Creative ignorance coupled with an optimistic mindset favors the opportunity to create a breakthrough innovation. This combination creates favorable conditions for what Frans Johansson called 'The Medici Effect' where breakthrough innovations occur at the intersection of disciplines, ideas and domains.

One other essential ingredient is courage. Courage is the ability and willingness to face danger and uncertainty. The path finders and path creators that have made a difference to all of our lives have consistently overcome their fears to keep walking toward a better future. They have consistently dreamed the impossible dream.

In the words of lyricist Joe Darion from the Musical *Man of La Mancha* they have sought 'To dream the impossible dream. To fight the unbeatable foe. To bear with unbearable sorrow. To run where the brave dare not go ... To try when your arms are too weary. To reach the unreachable star.'

I hope Professor Formica's timely book will inspire you to dream your impossible dream and reach the unreachable star.

Martin Curley
Vice President
Director, Intel Labs Europe Intel Corporation
Professor of Innovation, Maynooth University
August 2014

DOI: 10.1057/9781137492470.0002

Acknowledgments

Thank you, Martin Curley, distinguished leader of thought and innovation, for our long discussions on how to open up new doors and do new things in the current age of uncertainties.

Thank you, John Edmondson, master weaver of literary and economic discourses.

Thanks, too, to Daniel Crowley, shrewd observer and player of entrepreneurship in action, for additional research, and to Tim Feest for his extremely valuable contribution to the editing.

▶ Special thanks to Philip Nolan, President of the National University of Ireland (NUI) in Maynooth, and Ray O'Neill, Vice President for Innovation, for their active support and encouragement. I would also like to include in these thanks my colleagues Peter Robbins and Frank Devitt. Their guidance of the NUI Centre for Entrepreneurship, Design and Innovation (EDEN) is a constant source of inspiration.

Thanks are also due to Maurizio Guermandi & Associates for bringing the narrative to life with their line illustrations; to Nicola Farronato for allowing me to use the 'Rose of Emotions' illustration; and to the editorial and production staff at Palgrave Macmillan for getting the book into print and as an e-version.

In many different ways all of these people have enriched the book with their ingenuity.

The wind that blows on the islands of the Mediterranean evokes voices from the past that are projected into the future. Surrounded by silence, I was able to hear those

DOI: 10.1057/9781137492470.0003

voices as a guest in the Sardinian house of Amina Trevisani overlooking the windy Strait of Bonifacio. Thank you very much, Amina, for your outstanding hospitality.

DOI: 10.1057/9781137492470.0003

Note from the Author

'There is nothing permanent except change'—said Heraclitus of Ephesus, a pre-Socratic Greek philosopher.

Innovation is about change

What will the change be like? With the choices made today we choose the future we want. But it is not taken for granted that the future will be just as we imagined. The journey to the future is a true adventure during which we find uncertainty, the unknown, as our traveling companion. As Mark Twain says, with subtle irony, 'All you need in this life is ignorance and confidence, and success is sure.'

The majority of innovators think of change as an extrapolation of past events that seeks to improve what they already know how to plan and do. The probability of success can be measured: the risk lies in the measurement. However, committed to remake and invent a world that does not exist today, a minority of innovators decide to change themselves and leave behind the daylight of certainty, advancing into the dark night of unmeasurable uncertainty. For them, ignorance is not a barrier to action. What will happen along the pathway to the future they will discover as part of the process of remaking and inventing. The behavior of this minority seems to resonate with the lesson of the British economist John Maynard Keynes (1883–1946), who placed uncertainty at the heart of contemporary economic problems. Path creators, the

DOI: 10.1057/9781137492470.0004

minority leaders, are those who live in the certainty of uncertainty, and this is why they tackle the unpredictable head on.

According to tradition, firms innovate by responding to the customer's wants. The map of knowledge mastered by experts within a company and others who collaborate with them from the outside is the tool used to improve the performance of its products and services, and to win new customers. Once the path to be explored is drawn on the map, knowing how to get to the finish line is what determines the success or failure of incremental innovation.

However, more refined design and more efficient implementation are not absolute guarantees of success. Success will result from a paradigm shift in technology and business models, a disruption to the knowledge map. 'The Role of Creative Ignorance' is an essay in which creative ignorance enlightens non-experts to path creation, a phenomenon capable of dimming the brightest stars in the firmament of entrepreneurship. BlackBerry and Nokia have been damaged, hit by the assaults of Apple, Google and Samsung. The business models of Ikea and Zara have caused serious injuries to their Italian and other competitors.

Path creators have clever legs and a foolish brain. Their ideas are shaped by walking into the void of knowledge without predetermined destinations. They enjoy competitive pastimes, discover subtle analogies and, by sagacity and accident, exceed the boundaries set by knowledge maps to open up new, unprecedented routes and connect them to each other. In changing trajectories, path creators reveal latent, unexpressed needs of consumers, who will replace their traditional wants with new, revolutionary ones brought forth from seemingly nothing.

We do well to note the words of Carlos Castaneda (1970): 'Look at every path closely and deliberately, then ask this crucial question: Does this path have a heart? If it does, then the path is good. If it doesn't then it is of no use to us.'

Last but not least, I note that numerous facets of the two principal characters—the path finder and the path creator—are scattered throughout the narrative, forming a leitmotif of this literary opus. To give them resonance I have not always adhered to 'strict' scientific rules, in particular the dogma of prevalent economic thought. Remaining on the edges of the mainstream, the use made of ignorance is intentional.

DOI: 10.1057/9781137492470.0004

palgrave▸**pivot**

www.palgrave.com/pivot

Introduction: Knowing and Not Knowing

Formica, Piero. *The Role of Creative Ignorance: Portraits of Path Finders and Path Creators.* New York: Palgrave Macmillan, 2015. DOI: 10.1057/9781137492470.0005.

▶

In 1843, as he worked on his novel *Martin Chuzzlewit*, Charles Dickens wrote to his friend John Forster about the process of creativity:

> As to the way in which these characters have opened out, that is, to me, one of the most surprising processes of the mind in this sort of invention. Given what one knows, what one does not know springs up; and I am as absolutely certain of its being true, as I am of the law of gravitation....[1]

Dickens's surprise at his own capacity to develop the fictional characters that have attracted generation after generation of readers to his books is symptomatic of the creative dynamic which Professor Piero Formica takes as his subject in this volume. Dickens's short note to Forster tells us three things about radical creativity. First, it is a 'surprising process'—by implication, to allow it to happen, one must be open to surprise. Too much planning (that is, too much application of prior knowledge) stifles the unexpected; fear of the unknown is the enemy of creativity. Second, 'given what one knows, what one does not know springs up'. The inventor, Dickens, creates a character and puts it in a situation (what one knows), and then allows it to develop as a result of its interaction with its situation and with other characters, the creative process thus becoming unpredictable and the outcome as yet uncertain to the creator (what one does not know). Third, when the outcome is known, 'I am as absolutely certain of its being true, as I am of the law of gravitation'—the integrity of the creative process is assured because Dickens has allowed his creation to take its own path and find its own destiny. The outcome—a combination of the knowledge and skills applied to the initial creation or idea and the open-mindedness to and welcoming of the unknown in its development—is novel, arresting, enduring and true.

These critical elements of the creative process lie at the heart of Piero Formica's thesis in this book about the value of 'ignorance'—'what one does not know'—in invention, entrepreneurship and business development. His deconstruction of creativity through analogy, allusion and narrative becomes a testimony to the virtue and potential of embracing the unknown and the danger of being constrained by the known.

In his chapter 'Experimental Theatre', Piero Formica draws striking parallels between dramatic and business creativity. An actor acquaintance recently told me that each night on the stage was a process of discovery and learning, that he was constantly changing details of his performance in the same role, constantly reinventing his character as he found out what worked and what did not work, or as something new occurred to him as a result of some interaction with the other actors or between the performance and the audience on any given night. The demand for creativity, in other words, was permanent and he never knew what new idea that night's performance would reveal to him.

The stage, or 'The Empty Space' as Peter Brook called it,[2] provides a strong metaphor for the creative process. Faced with a set of bare boards and with the knowledge of what has been done with it before, the director finds new ways of presenting old plays, the actor finds new ways of performing old roles, the radically creative playwright introduces new forms of theatre. The 'live' nature of theatre makes it always unpredictable, always changing, and the playwrights, actors and directors who have had the greatest impact have been those who have taken the greatest risks and who have been able to marry their expertise to a courageous openness to the inherent unpredictability of the art—combining, in other words, 'what one knows' and 'what one does not know'. Let's consider three examples of theatrical innovation from, respectively, a director's, an actor's and a dramatist's perspective.

In directing, Peter Brook's famous 1970 production of *A Midsummer Night's Dream* effectively reinvented one of Shakespeare's most frequently performed plays, giving actors and audiences alike an entirely new perspective on it. Writing in 2013, Brook described how the radically different production took shape:

> So we began with only the conviction that if we worked long, hard and joyfully on all the aspects of the play, a form would gradually appear. We started preparing the ground to give this form a chance. Within each day we improvised the characters and the story, practised acrobatics and then passing from the body to the mind, discussed and analysed the text line by line, with no idea of where this was leading us. There was no chaos, only a firm guide, the sense of an unknown form calling us to continue.[3]

This is not mystical gobbledygook about the creative muse. Brook stresses that he and his group of actors brought all their experience to the experimentation, but the critical element in the development of the innovation was an openness to the unknown. 'If the concept is imposed in advance by a dominating mind,' writes Brook, 'it closes all the doors.'

In acting, Anthony Sher's 1984 radical re-interpretation of Richard III, playing the character on crutches, broke the mould of Laurence Olivier's 1955 interpretation which had remained the definitive one until Sher's experiment found a new direction. Sher recalls that his startlingly novel

DOI: 10.1057/9781137492470.0005

re-creation of the 'bottled spider' was driven by a conscious need to innovate:

> When I played Richard, I did a lot of research—I was trying to find a way of inspiring myself to play a role that had so famously been played by [Laurence] Olivier.... Later, I would discover that all Shakespeare's great roles have had famous interpretations by famous actors, and this is just part of the job. But at that time I was obsessed with trying to find my own way of playing Richard.[4]

The evolution of Sher's acclaimed performance involved painstaking research, discussion and experimentation—the challenge was to produce an innovation that at once rejected and matched the hugely successful 'incumbent', Laurence Olivier's world-renowned and much imitated performance.[5]

In writing, Samuel Beckett's 1972 disturbing and unforgettable *Not I* broke all the rules. The play is a roughly 20-minute high-speed monologue spoken in darkness by an actress, unseen except for her mouth, a few feet above the stage and illuminated only by a spotlight. With almost everything normally associated with a theatrical performance absent, *Not I* is a venture into the unknown via a deliberate rejection of the known. Anticipating rehearsals for the 1973 performance by Billie Whitelaw at the Royal Court Theatre in London, Beckett wrote 'Rehearsals in London December. Hope to find out then if it's theatre or not.'[6] Openness to the risk of failure goes hand in hand with radical creativity.

For Beckett, and for Dickens, Brook and Sher too, ignorance of what would result, of where their ideas would lead them and openness to the unknown were essential elements in the development of their innovations. The same is true, as Piero Formica shows us in the pages that follow, for path-creating entrepreneurs and radical innovators in business, technology and science. This book is an exploration of what is and always has been the driver of progress and positive change—the creative mind.

John Edmondson
Director, IP Publishing Ltd
London, August 2014

Notes

1   Charles Dickens, letter to John Forster of 1843, in Madeline House, Graham Storey and Kathleen Tillotson, eds. (1974), *The Letters of Charles Dickens*,

DOI: 10.1057/9781137492470.0005

Volume 3, 1842–1843, Pilgrim Edition, Oxford: Clarendon Press, p. 441. Dickens's insights into his own creative processes are discussed and extended by Barbara Hardy (2008), in *Dickens and Creativity*, London and New York: Continuum.

2 'I can take any empty space and call it a bare stage.' Peter Brook (1968), *The Empty Space*, London: MacGibbon and Kee.

3 'Peter Brook on A Midsummer Night's Dream: A cook and a concept', *The Guardian*, 15 April, 2013.

4 Anthony Sher, in 'Richard III: Shakespearean actors rake over the remains', *The Guardian*, 4 February, 2013.

5 Anthony Sher gives a detailed account of the evolution and development of his Richard III in *Year of the King—an Actor's Diary and Sketchbook*, Pompton Plains, NJ: Limelight Editions, 2006.

6 Quoted in Deirdre Bair (1990), *Samuel Beckett: A Biography*, London: Vintage, p. 666.

DOI: 10.1057/9781137492470.0005

Part I
The Role of Creative Ignorance as a Willing Action

▶

DOI: 10.1057/9781137492470.0006

1

Why Is Creative Ignorance Important?

Formica, Piero. *The Role of Creative Ignorance: Portraits of Pathfinders and Path Creators*. New York: Palgrave Macmillan, 2015. DOI: 10.1057/9781137492470.0007.

▶

Knowledge   Creative
                             ignorance

DOI: 10.1057/9781137492470.0006

With good reason, ignorance is generally ill-regarded. However, there are opportunities for observing ignorance from other perspectives, brought to light not only by researchers but also by the popular media. Introducing, on 17 August 2014, the BBC Radio 4 programme 'Something Understood', which addresses each week a different and wide range of topics under the guidance of Mark Tully, a long-serving and highly experienced BBC journalist, the radio announcer said that in this particular programme, 'On Ignorance',

> Mark Tully invites us to accept our own ignorance as a first step on a voyage of discovery, taking his lead from Socrates' well-known thought that, 'The only true wisdom is in knowing you know nothing.'

There is a lack of awareness of creative ignorance, that which by design comes after, not before, knowledge and unlocks otherwise unthinkable paths of economic growth and social development, and is even less well-known for being revolutionary. As humanity struggles supposedly to eradicate ignorance, those who lobby for knowledge and expert groups push creative ignorance into a corner. It was not thus in the ancient world. Scholars and wise individuals of times past were those who pioneered the idea of creative ignorance. In today's world, we need more than ever 'Homines Novi', New Men, who enable us to exploit the strengths of creative ignorance and overcome the weakness of accrued knowledge.

Creative ignorance contains a good dose of unreasonableness. At the end of the 1980s Charles Handy, acclaimed scholar of management, wrote that,

> We are entering an Age of Unreason, when the future, in so many areas, is there to be shaped, by us and for us; a time when the only prediction that will hold true is that no predictions will hold true; a time, therefore, for bolding imaginings in private life as well as public, for thinking the unlikely and doing the unreasonable. (Handy, 1989)

Equally, others have pondered on the wisdom of ignorance:

> Our wretched species is so made that those who walk on the well-trodden path always throw stones at those who are showing a new road. (Voltaire, Philosophical Dictionary)

> I am looking for a lot of people who have an infinite capacity to not know what can't be done. (Henry Ford)

DOI: 10.1057/9781137492470.0007

The *adventum*

Engaged in scientific research or involved in the world of business, what decisions do we take once we reach the frontiers of knowledge? Do we attempt to overcome the boundaries of incomprehension by using acquired knowledge and past experiences? Stripped of the baggage of our existing knowledge, do we continue the quest for new knowledge by embracing creative ignorance? Do we embark upon a project that takes us away from John Milton, to follow in the footsteps of William Shakespeare as interpreted by William Hazlitt (1778–1830)? This English literary critic and essayist wrote,

> Uneducated people have most exuberance of invention and the greatest free-dom from prejudice. Shakespeare's was evidently an uneducated mind, both in the freshness of his imagination and the variety of his views; as Milton's was scholastic, in the texture both of his thoughts and feelings....If we wish to know the force of human genius we should read Shakespeare. (Hazlitt, 1822)

Or, perhaps, do we invoke God or deities, as scientists of the highest calibre such as Isaac Newton, Pierre-Simon de Laplace and Christian Huygens sometimes did? Neil deGrasse Tyson (2005) captured this nicely in recalling the invocation of Newton to God:

> Eternal and Infinite, Omnipotent and Omniscient; ... he governs all things, and knows all things that are or can be done. ... We know him only by his most wise and excellent contrivances of things, and final causes; we admire him for his perfections; but we reverence and adore him on account of his dominion.

There are many questions preying on our minds when we embark on the passage of discovery of paths to tread. In the course of our adventure, what might befall us, or what we want to happen? Will our *adventum* ('venture') be as we had supposed? Will it be wider, narrower or more different than we ever thought imaginable? To answer these questions, each explorer will use the tools available in their tool kit. According to attitudes, motivations, capabilities and circumstances, it may be the tool kit of knowledge or that of creative ignorance. There are those who want to master the attributes of knowledge at the highest level in order to find the right direction. There are others who, like Socrates, have learned that ignorance is the very thing that makes them wiser than the others, that gives them the capacity to make creative decisions about where and

DOI: 10.1057/9781137492470.0007

when to go. In times of major transformation, this second species of individuals is suddenly revealed, appearing almost as a major mutation of human behavior. These, the 'hopeful monsters' as Goldschmidt (1940) called them—albeit in a completely different context—bring the hope and promise of changing the rules of the game. In doing so, they would plot a set of trajectories quite unlike those that had previously prevailed and would give voice to new players. When the 'monsters' appear, the Very Knowledgeable Persons desperately seek support in someone who, like Dante's Virgil, possesses the reason and decisiveness required to repel their attempts to create the conditions for the birth of a new world order.

It seems somewhat obvious that there is a direct and linear channel of communication between knowledge and ignorance—a feeling like that of a pair of tango dancers. Where the one is lacking, the other appears. When I know that I don't know, then I rally my energies to close the gap. This is in fact what researchers and entrepreneurs do when they reach, respectively, the borders of their scientific and entrepreneurial knowledge. However, matters begin to attain both a higher level of difficulty (they become more complicated) and complexity (a greater number of components comes into play) once two causes arise, together or separately. The first cause is the 'not knowing of not knowing'. That one may not have a mind open enough to be able to get rid of past experiences and biased reasoning, and which is not affected by and vulnerable to external criticism, as well as being ill-informed, must also all be taken into account.

It can also happen that closed-mindedness can have such an influence as to cause us to exchange one thing for another. As described in *The Little Prince* by Antoine de Saint Exupéry (1943), the boa constrictor mistaken for a hat is a sign of unimaginative minds for which the only possible explanation is their limited perspective. The protagonists of *The Little Prince* make us appreciate a child's mind full of imagination and empty of those past experiences that are the sources of stereotypes and prejudices. Exupéry points at grown-ups as having narrow minds. What is certain is that for them the journey from knowledge to creative ignorance is a metaphorical walk on hot coals: they then come to see that knowledge has more or less relevance depending on who is its bearer. Returning to the story of The Little Prince, the Turkish astronomer who, at the International Astronomical Congress in 1909 presents with some brilliance his discovery of an asteroid, is ignored by the audience of scientists,

DOI: 10.1057/9781137492470.0007

for he was wearing the traditional clothes of his country; and this is why 'A Turkish dictator made a law that his subjects, under pain of death, should change to European costume. So in 1920 the astronomer gave his demonstration all over again, dressed with impressive style and elegance. And this time everybody accepted his report' (de Saint Exupéry, 1943, Chapter IV). Because fantasy and reality are mirror images of each other, the episode of the Exupéry's Turkish scientist has been replicated in the reality of academic life a thousand times and more. We find evidence of this in academic elitism and hubris ('unchecked intellectual arrogance'), as highlighted in all their harshness by Professor Martin Anderson (1992): two mindsets that give so much credibility to the knowledge of their bearers. In addition, we must consider the fact that knowledge is naturally inclined to search for errors with a view to removing them, making use of analysis, investigation and expertise. Creative ignorance, for its part, constantly searches for the inner nature of things through intuition.

Let us imagine for a moment that we can build a box that includes all of our own knowledge, beliefs, habits, ideas, customs and social behavior, but not ignorance. We call this artefact 'Culture'—that is, our personality or 'human operating system', as defined by Bill Tobin, a consultant at Strayer Group in Silicon Valley. The code of personality rejects ignorance. Knowledge is light; ignorance is opacity and darkness. In ancient times, the Egyptians worshiped Ptah, the god of knowledge and learning, of creation, the arts and fertility. In the epistolary novel 'Augustus' (Williams, 1972), the American author John Williams mentions Athens as 'the mother city of all knowledge' in his imaginative creation of the Roman poet Horace's thoughts and stages 'an Italian Orpheus [whose] love was no woman; his Eurydice was knowledge'. Philostratus, a Greek sophist of the Roman imperial period, in his book *The Life of Apollonius of Tyana* (Jones, 2005) ascribed to gods, as possessors of all knowledge, the ability to pierce the veil that covers the future. And in Dante's *Inferno* ignorance is, together with impotence and hatred, one of the three characteristics attributed to Beelzebub, one of the seven princes of Hell. In short, without knowledge there are no paths that we can find or create, and then tread. As stated in the poem Ithaca by Constantine Cavafy (2007),

> When you set out on the journey to Ithaca, pray that the road be long, full of adventures, full of knowledge.

But, consider this: if lacking a particular type of ignorance were not possible, would that make us feel quite adventurous?

DOI: 10.1057/9781137492470.0007

The way of treating ignorance in early times was to separate the ignorance of the foolish from that type of a healthy Socratic, 'self-aware' kind of ignorance which Saint Augustine dubbed as mindful, learned ignorance. Down the centuries that followed such a concept was reaffirmed by the humanist, mathematician and astronomer Nicolaus Cusanus (1401–1464) in his work *De Docta Ignorantia* (On Learned Ignorance) where, in the words of Philippe Verdoux (2009), he held that '… ignorance and knowledge are not wholly distinct epistemic phenomena, but combine and overlap in interesting ways … . The more [a wise person] knows that he is unknowing, the more learned he will be. In other words, learned ignorance is not altogether ignorance, but a kind of knowledge or wisdom'; by the German philosopher Johann Gottlieb Fichte (1762–1814), who argued that down the road towards the conquest of 'not knowing' is an infinite journey; and by the American educational reformer John Dewey (1859–1952) who explained that there is a type of ignorance—he called it 'genuine'—which is 'profitable because it is likely to be accompanied by humility, curiosity and open-mindedness' (Dewey, 1933).

In our lifetime, Hans Magnus Enzensberger, one of Europe's leading writers and critical thinkers, states that the acquisition of knowledge requires gestures of refusal. In order to see something you have to give up many things to see. Ultimately, we start from a position of ignorance to create new branches of knowledge beyond, or to replace, those that already exist. Those who take this course of action are aware of the extent of their ignorance: in the words of Confucius, this awareness is their true knowledge. With existing knowledge, one keeps delving far more deeply into the same pit. In the Internet age, a shroud of mist obscures learned ignorance—that is, the creativity from which all things new emanate. Thanks to the Internet, it seems that we are all knowledgeable, confusing facts and figures (information) with cognition (knowledge) as a result. Even when information and knowledge are kept apart, the latter is the breath of our intellectual life of such force as to stifle creative ignorance.

The Philosopher's Path and the tightrope walker

Knowledge gained enables suitable ways to pursue an approach to incremental innovation to be found. The smart path to take looks like The Philosopher's Path in the northern part of Kyoto's Higashiyama district.

DOI: 10.1057/9781137492470.0007

This is the path of meditation, with flowering cherry trees in spring-time. Nishida Kitaro (1870–1945), a prominent Japanese philosopher, practised meditation while walking this safe and enjoyable route. After the blossom comes the harvest, not always equally abundant. Similarly, the incremental approach to innovation prompts meditation to find methods that lead to blossoming and harvests better and more abundant year after year. Ultimately, however, it is one and the same topic: cherry trees along The Philosopher's Path. The repetitions reinforce the rules and strengthen the order, and the recurring motives cushion the blows of initial surprise.

Let us now look at the learned ignorant in their creative role. A tight-rope walker in mid-air over two skyscrapers has replaced the philosopher musing on the path of cherry trees. Our funambulist has the intellectual stance of Philippe Petit, an inimitable high-wire artist who, in his own words, has 'learned to welcome life's surprise, cheating the impossible, disregarding the rules, and learning to unknot the problem (I'm tempted to say "the streetwise way") as opposed to focusing on acquisition of the right answer—possibly one major flaw of what I would refer to as a formal education' (Petit, 2012). From the tower where the exercise starts—the knowledge skyscraper—the many people who live there are watching the acrobat moving towards the half-empty skyscraper of creative ignorance. Will the performer reach that building safely? The risk is high, because of the strong winds that blow in the direction of the endpoint. These are the winds of a cultural bias because, however you would like to describe it, together with the reasoning promulgated by conventional wisdom, ignorance is always thus: it does not go hand in hand with the adjective 'creative'. Knowledgeable individuals, being careful and thorough, are firmly convinced that, due to his ignorance, the acrobat has overstated his capabilities. In other words our tightrope walker, being ignorant, does not possess the skills needed to recognize their limits. Precisely the opposite is the case: the fact is that the knowledgeable persons are victims of a weak extension of their ignorance.

What emerges from this metaphor is that creative ignorance is a source of uncertainty. In spite of this, the pace of a path creator is not slowed by the uncertain consequences that they must face: quite the contrary, in fact, for the pace is accelerated by the motivations that led them to proceed. Path creators prefer to sail in the unknown waters of uncertainty rather than to linger comfortably in the pond of certain knowledge, which only gives a sense of false security. Following the train of thought developed

DOI: 10.1057/9781137492470.0007

by John Maynard Keynes in his *General Theory*, where uncertainty seems to be equated with a normal situation (Skidelsky, 1992) it is the ethics of human motivation and not those founded on the consequences of their actions which calls upon the creators to take action.

The knowledge skyscraper is inhabited by the experts—those who show 'special skill or knowledge because of what [they] have been taught or what [they] have experienced', as defined by the Merriam-Webster dictionary. Knowledge is a word compromised too much by its links with these experts; so, more often than we would like, it happens that the rules of the knowledge game force the expert to be focused on planning, process and compliance. In short, the expert manages the stages of replication and improvement. The skyscraper of destination—that of creative ignorance—is, unlike the knowledge tower, almost uninhabited, inasmuch as ignorance has a bad reputation and arouses fear even when it has the touch of a creative mind. Creative ignorance encompasses events with a high degree of uniqueness and of unmeasurable uncertainty. The creative ignorant, focused on observation and curiosity for change, lays down new, unprecedented paths—for invention, innovation or entrepreneurship.

This kind of event is far removed from that of those who walk the usual paths or identify others by making use of their tested experience. Competent people argue that ignorance, no matter how creative it may be, gives rise to major risks of failure; or, at best, success is very uncertain and no one knows when it will occur. This is why the experts do not exceed the bounds of their knowledge and why investors are not willing to take big risks by deciding to get ahead, beyond the *finis terrae* of knowledge, the limit of their sight. Within that boundary competent people sail to a charted land.

Scientists argue that we live in a quantum society—'qbit'—and as such we would be like particles whose position, time and impact on other particles, and vice versa, are uncertain because of continuous technological upheavals. We are also immersed in the universe of information. 'It from bit'—stated otherwise, everything is just made of bits—is the phrase coined by John Archibald Wheeler (1911–2008), an American theoretical physicist who worked with Niels Bohr. In order to give sense to information, we have to complement it with knowledge—that is, the deeper understanding of how things work. To do this, we dig deep wells that we call 'specialization'—exemplified, for instance, in the fragmentation of science into dozens of specialized components. As Ronald Wright

DOI: 10.1057/9781137492470.0007

noted in his book *A Short History of Progress* (Wright, 2004), specialists are 'people who know more and more about less and less, until they know everything about nothing'. The deeper the well, the less the light that penetrates into it. Each specialization is a repository of knowledge, better known in the professional circles as the 'knowledge map', where nothing should be left to chance and nothing improvised—something that is perhaps reassuring but which at the same time exposes its outer edges to dramatic events. We need only to recall the sinking of RMS Titanic in 1912, a vessel of which it was said at the time that nothing had been left to chance according to the knowledge of experts, who failed to recognize their errors. The sinking of that giant of the seas was perhaps as much a result of the very nature of the frame of mind that character-ized the experts' working style as it was the iceberg.

There is a moment in time when the truth of a deep but narrow vision is no longer certain. It is then that the knowledge map used to dig the well should be abandoned and replaced with creative ignorance, which by design comes not before but after knowledge and sets up contacts with the unknown. Knowledge can be deceptive, whereas the astuteness of creative ignorance unlocks otherwise unthinkable paths of economic growth and social development. The stronger the tension, the more crea-tive ignorance moves away from the knowledge gained, to set out into uncharted territories that in the future will become familiar to everyone. This means that today's creative ignorance is a call to tomorrow's knowl-edge. It often happens that the creative ignorant are outsiders—that is, non-experts who, not knowing a subject, tackle it from a totally different angle than the seasoned experts who know it thoroughly, or those who no longer feel bound to share the ideas of the master who drew that map.

Creative ignorance

Knowledge and creative ignorance are the two leading trendsetters of innovation. One day, innovation wears clothes that knowledge designs: then it is the turn of creative ignorance to dictate the fashion. What happens in one scenario and then the other? Being familiar with the market, knowledgeability hastens incremental innovation—that which makes good products better for consum-ers who want improved versions of the products available in the markets. Knowledgeable entrepreneurs are subjected to rational

DOI: 10.1057/9781137492470.0007

ignorance, which occurs when the cost of exit from their market-bounded knowledge maps and entry to the field of creative ignorance is deemed of greater value than the intended benefits from choosing the latter. The creative ignorant who, staying away from the market, give rise to novelty and diversity that confound the existing markets, escape from this syndrome.

Creative ignorance promises to constitute a significant shift for achieving technological changes as well as changes in organizational and societal models that are a novelty. So, we can gain access to new levels of knowledge, as happened with electricity which, for instance, made it possible by using electric lighting to lower the ceilings in buildings and therefore enable more floors to be installed (in our case, new floors of knowledge) for the same height of a building.

Lacking in creative ignorance, the future would be even further from us, as was the ceiling from the floor, due to problems of soot and overheating, in rooms lit with candles and oil lamps.

In its work, creative ignorance proves to be

▸ Conscious—Aware of one's ignorance and responding to one's surroundings;
▸ Intentional—Done with intention or on purpose;
▸ Insightful—The act or outcome of grasping the inward or hidden nature of things or of perceiving in an intuitive manner and
▸ Perceptive—Able to see what others cannot.

When it comes to acting for the future and having the power to make it happen, most people recognize that they are replete with past knowledge but devoid of the creative ignorance that 'follows knowledge' (Firestein, 2012), supersedes the content that stays in the body of knowledge and has built up over time, and is coupled with serendipitous discoveries. But when does the time come? The time is that of Schumpeter's creative destruction and Christensen's disruptive innovation of businesses which, until recently, seemed so solid as to be unassailable. This is the time when *Homines Novi*, the 'New Men', enter the picture—people who did not immediately assume that something radically new could not be accomplished. Paraphrasing 'Etiam si omnes...ego non' from Saint Matthew's Gospel (26:33, 35), 'Even if all others follow the knowledge, we will not', *Homines Novi* do not abandon it [creativity] which, in contrast, eludes the captors of knowledge, and add to it conviction, without which the path on which they have set out would not be viable. Among them there

DOI: 10.1057/9781137492470.0007

are the disciples of the three Princes of Serendip (Merton and Barber, 1992)—travellers taking a trip into Innoland, the island of innovation (Formica, 2013), and stumbling on something they were not looking for but recognizing its significance immediately. It has been estimated that thirteen times more successes than failures arise from the 'accidental sagacity' of these random events (Franklin, 2013).

Ignoring or overlooking today's consumers for creating the consumers of tomorrow

With a view to Clayton Christensen's theory of disruptive innovation (Christensen, 1997), starting from scratch (the non-consumption area), disruptors create new consumers. At the very beginning, their products look like 'ugly ducklings', because their performance fails to meet the current standards.

From the perspective of the serial entrepreneur Steve Blank, disruptive innovation

Is the innovation we associate with start-ups. This type of innovation creates new products or new services that did not exist before. It's the automobile in the 1910s, radio in the 1920s, television in the 1950s, the integrated circuit in the 1960s, the fax machine in the 1970s, personal computers in the 1980s, the Internet in the 1990s, and the Smartphone, human genome sequencing, and even fracking in this decade. These innovations are exactly what Schumpeter and Christensen were talking about. They create new industries and destroy existing ones. And interestingly, in spite of all their resources, large companies are responsible for very, very few disruptive innovations (Blank, 2014).

DOI: 10.1057/9781137492470.0007

2
Introducing Path Finders and Path Creators

Formica, Piero. *The Role of Creative Ignorance: Portraits of Path Finders and Path Creators.* New York: Palgrave Macmillan, 2015. DOI: 10.1057/9781137492470.0008.

▶

DOI: 10.1057/9781137492470.0008

'Path Finder' and 'Path Creator': two portraits, two characters that appear in caricature more pronounced than may be the case in living reality. Reasonable and rational persons, path finders seek and search by reviewing the knowledge maps they hold and behaving accordingly. They are consistent with the world of knowing. In contrast, path creation occurs through a learning process of creative performance that is at complete odds with knowing. Discarding any form of knowledge map, path creators go beyond the limits of reason, to penetrate the 'not-knowing' space. Whereas the finders face risk when they decide to set out in search of a path within their maps, the creators live the uncertainty and unpredictability of those who create a new path from nothing. They do not resort to knowledge practices, rules and handbooks.

What is even more valuable is that path creators are willing to act rather than being fuelled by the ambition of what one should do and how to do it. It is a willingness that relies on intuition and judgment and which makes use of multiple alternatives, rules of thumb, random processes and field experiments. Sifting through their behavioural baggage, one searches in vain for rational expectations, defined by Frydman and Golberg (2007) as an ideology that leads to the construction of models of reality that are exactly wrong.

> Inventing the future requires giving up control. No one with a compelling purpose and a great vision knows how it will be achieved. One has to be willing to follow an unknown path, allowing the road to take you where it will. Surprise, serendipity, uncertainty and the unexpected are guaranteed on the way to the future. (George Land)
>
> Knowing becomes the antithesis of learning. (George Land)
>
> I don't believe in learning from other people's pictures. I think you should learn from your own interior vision of things and discover, as I say, innocently, as though there had never been anybody. (Orson Welles)

The Hamlet dilemma: To be a path finder or a path creator?

The path of economic growth is traced in the territory of macroeconomics. It is the direction in which the paths of microeconomic territories move together. These latter include the paths of innovation followed by the path finders and

DOI: 10.1057/9781137492470.0008

those invented by the path creators. Relying on knowledge of market data and promptly alerted when seeking new ones, path finders reinforce the existing rules of the game. In doing so they contribute to GDP growth in the short to medium term. Path creators start from a condition of zero, a kind of *tabula rasa* ('blank slate'), which causes them to revolutionize the rules of the game by changing substantially the sources of economic growth in the long term. Think of the computer scientist Tim Berners-Lee who created the World Wide Web path and of the artisan John Harrison who, in the second half of 1700s and encouraged by the prospect of prize money worth about £3m today, created a path which made possible more accurate, and thus safer, navigation. The question remains open as to where the balance between path finders and path creators lies that best helps in keeping to sustainable paths of growth.

We choose and make decisions every day thanks to the path charted by William Shakespeare. Looking out of the window in search of emotions, choices force us to reflect on who we are and ask who we want to be in the future. 'To be or not to be' is the name of the path that has brought humanity into the modern age. Who treads that path makes a choice, and once the initiative has been taken, meets the challenges of its consequences. Like Hamlet, innovators must decide which path to walk. To be a path finder or a path creator? That indeed is the question. To make forecasts for the future with the tools of knowledge and then, as Joi Ito, the head of the MIT Media Lab, says (Ito, 2014), stay on 'hold' until you have proof that your idea is conforms to the expectation? Or choose to replace forecasts and expectations for immediate action, 'being open and alert to what's going on around you right now'? In short, to be a 'now-ist' path creator?

The innovation rush is underway, spanning all physical continents of planet Earth and all conceptual continents of planet Discovery. Future and current entrepreneurs from generations X (the post-World War II baby-boomers), Y (the Millennials) and Z (the youngest cohort of people born from mid- to late 1990s) take part in this feverish competition. A generally buoyant feeling of 'being the winner' motivates each individual contestant. From this perspective the innovation rush has similarities with the gold rushes in the nineteenth century. It is a race that provides a stimulus for migration and international movement of talented people, opens up new commercial routes and urges cities to seize the crossroads either of those paths most followed or those showing most promise.

The Romans used the verb *patēre* to indicate 'I am open', 'I am accessible', 'I am exposed'. The word 'path' can be related to this verb, but also

DOI: 10.1057/9781137492470.0008

to the Latin noun *limes*, which denotes a demarcated field, a boundary line. Therefore, there are paths indicated by a frontier and paths in the form of open spaces along which the wayfarers create or go in search of opportunities, and find solutions to difficult or convoluted problems. It is not for nothing in 1956, when Ray-Ban introduced a revolutionary design in sunglasses, that the new model was named Wayfarer.

There are those who follow a path in pursuit of innovative and valuable ideas and whose profile matches either that of the path finder or the path creator. Their portraits, that we are going to draw, reflect the view of the Italian poet Giacomo Leopardi (1798–1837):

> A portrait, even if a very good likeness (indeed, especially when it is such), not only tends to have a greater impact upon us than the person depicted (which comes from the surprise that derives from imitation, and the pleasure that comes from the surprise), but, if I may so put it, that same person has a greater impact upon us painted than real, and we find them more beautiful if they are beautiful, or conversely, etc. If for no other reason than because when seeing that person, we see them in a ordinary way, and seeing the portrait, we see the person in an extraordinary way, which increases unbelievably the acuteness of our mind and faculties, and generally greatly enhances our sensations, etc. (Leopardi, 2013)

The two portraits show us two different characters, the path finder and the path creator, who travel beyond today's places by striking out on new paths. As did the Arthurian hero Percival, the two personas undertake a process of self-healing from past behaviors and conceptions in their quest for the 'Holy Grail' of innovation. The portraits highlight the expansion of their consciousness to observe events from a new perspective and understand and address the questions that matter, and enables them to shed new light on actual wants (the path finders) and unveil hidden needs (the path creators).

Two couples: wants and old knowledge, hidden needs and creative ignorance

It is the case that we might be condemned by facts that come from information and then are processed by knowledge. It is also the case, in contrast, that we might be rewarded by ignorance that, by its nature, is devoid of facts. To use the words of Nicholas Rescher (2009), 'With ignorance we do not have the facts.' In the first case one is vulnerable to inconsistencies and misinterpretation of the facts, or that the facts of today do not create the reality of

DOI: 10.1057/9781137492470.0008

tomorrow. In the second case, one might benefit from not being biased by facts that affect choices: there are no facts to cast a shadow on what is done.

In the Knowing Land the wants are the large front door that gives free access to the facts that happen in the markets, with producers who, as market readers, have a good command of their knowledge maps, are focused on their strongest 'knows' and surrounded by experts who supplement their 'knows', scrutinize the demands of consumers, capture the consumer's 'wants' and meet specific requirements of the most faithful customers. The hidden needs are just a small side entrance, almost invisible and made inaccessible. Reading through the lenses of old knowledge it is not necessarily the case that facts reflect reality in the making. The old knowledge hinders the formation of new knowledge. It is the search of creative ignorance, through learning how to cultivate what is called a 'beginner's mind—a mind that is willing to see everything as if for the first time' (Gelatt Partners, 2008), that lifts the veil of knowledge. Removing this veil, which prevents us from seeing the true reality as it unfolds, we will not be deceived by the illusion of having grasped that reality. If the facts mislead, even to the extent of becoming opponents of the truth, there is nothing we can do other than try to rely on creative ignorance—that which causes us to have doubts—and, at the same time, ask ourselves tough questions. This is how to go through the door of hidden needs and to turn them into products and services available for the first time in the market.

All other things being equal, those going through that door are the seekers of need who, having set their imagination in motion to go everywhere, dismiss ideas that are purely practical (i.e., tried and trusted) in favour of creative alternatives—novelties which trigger feelings of uncertainty. With a Zen attitude they use their beginner's mind and find themselves confronted with a wide range of fresh possibilities, in contrast to those experts sunk in their well of knowledge. Together with non-experts—integrative thinkers, who practise creative ignorance—some of these beginners, continuing to climb the steps of the ladder of success, end up becoming bored and, therefore, aspire to do things very differently. The same also applies to the group of people who, having reached the top of the ladder of success, are not afraid of walking quickly down the stairs to start soon after another climb far more challenging than the previous one. Last but not least, there are people who are at the foot of a steep climb to success and don't know if they are up to the challenge they face. Highly motivated, they do not fear uncertainty and, disregard the risk as something alien to their mindset:

DOI: 10.1057/9781137492470.0008

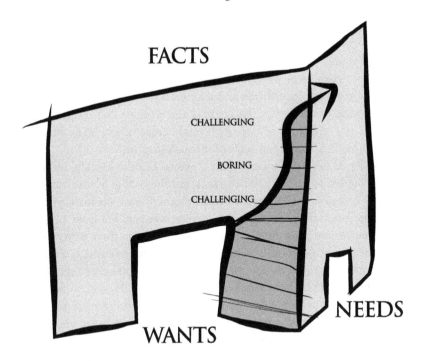

Reasonable and unreasonable people

Path finders and path creators live in separate habitats. To paraphrase George Bernard Shaw, path finders are reasonable persons who adapt the evolution of their thought to their knowledge maps. The first duty of path finders is to deliver results, often short-term ones, and therefore they are forced to stay in the mindset of 'knowing'. In contrast, path creators are 'unreasonable' persons who cause fundamental alterations in the previously mastered knowledge map, to the point of abandoning it. Through the gesture of rejecting the knowledge map, path creators refuse to see many things, attempting instead to go somewhere else in such a way that a profound renewal can take place. This means that they thrive on the challenge of the unknown and are curious and eager to see what happens once they move away from the knowledge map to occupy the space of 'not knowing'.

'There's no certainty of tomorrow', exclaims Lorenzo the Magnificent (*Lorenzo il Magnifico*), banker and ruler of the Florentine Republic in Renaissance during the fifteenth century, in his *Song of Bacchus*; and John Maynard Keynes—'the saviour of capitalism in its darkest hour' in the

DOI: 10.1057/9781137492470.0008

words of Robert Skidelsky—firmly believed that market participants face the irreducible uncertainty of future events. Accordingly, the logic of choice under conditions of uncertainty is the major factor behind the behavior of path creators. Looking forward, to see what could possibly happen, they exalt incertitude and unpredictability and regard them as integral parts of creativity. The sense of uncertainty that has dotted the history of Italy since the Middle Ages might create a potential source of a new Renaissance under the aegis of the path creators.

To summarize: path finders, who move within the borders of knowledgeability, see what everyone else sees, but they come to think and plan what no one else has thought of and designed. Path creators, who run across the boundless prairies of creative ignorance, see what others do not perceive and gain revolutionary insights from their observations. From their standpoint, what matters is that the actions which have been taken are valuable, even when they do not lead to the expected result.

But there is a third habitat, which the scientists call the 'ecotone', where the two habitats overlap into one another and tension arises between the risk of the deductions that the path finder has made and the uncertainty of the path creator's hunches. Both set in motion a sequence of actions for which we may use terms such as experimentation and exploration.

Path finders and path creators put together under the 'S-curve' rubric

The S-curve cycle of creation, improvement and replacement of innovations (Rogers, 2003; Geroski, 2003) is also a story about path finding and path creation. Many technology advances follow an S-curve (sigmoidal) process of gestation, growth, maturity and obsolescence (Figure 2.1).

In the early stages of the development of a new technology, ideas abound but no clearly dominant design or approach emerges. Radio, automotive and aviation technologies in the 1900s and 1910s, personal computers in the 1980s, and web technologies in the 1990s were all radically new technologies with, at the time, unclear futures and widely varying initial designs. As these industries matured, a small number of dominant designs emerged which were widely adopted. In all cases a long period of incremental innovation followed, typically very lucrative for industry incumbents who could make constant incremental changes and yet rely on a stable fundamental design and business model. In many

DOI: 10.1057/9781137492470.0008

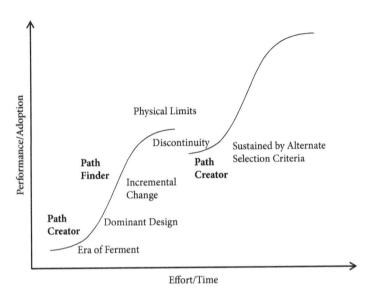

FIGURE 2.1   *Technology Lifecycle S-curve.*

cases, these designs ultimately were replaced by new discontinuities—jet aircraft for piston-engine propulsion in aviation; mobile and tablet computing for PCs in personal computing. In automotive engineering it is only now that we can see commercially viable electric/hybrid-powered vehicles, car sharing and, ultimately, driverless cars, bringing an end to a century-long dominance of the business models of Ford and General Motors.

The evolution of radiowave transmission and reception from an unknown, radical technology in the late nineteenth and early twentieth centuries to a mature technology in the 1930s and 1940s is characteristic of this curve. Early innovators were path creators who eventually established the dominant designs that resulted in the widespread adoption of radio in the home. As the technology matured further, incremental improvements came from path finders navigating this particular existing technology landscape.

In time these advances began to exhibit diminishing returns—for example, an analogue, vacuum-tube-based desktop radio could only be improved so much. New path creators, thinking about personal radio and music listening, turned their attention to creating an entirely new platform, made possible by fundamental advances in the miniaturization of electronics by means of the transistor and the integrated circuit.

DOI: 10.1057/9781137492470.0008

Miniaturized tape players like the Sony Walkman' and the personal transistor radio revolutionized both the radio manufacturing industry and also related industries such as music, advertising and entertainment. Path creators created these opportunities. Path finders then navigated and sustained this growth until, as opportunities involving the existing technology became limited, the time was right for a new generation of path creators to begin again the cycle of destruction and creation.

DOI: 10.1057/9781137492470.0008

3

The Embrace between Science and Entrepreneurship

Formica, Piero. *The Role of Creative Ignorance: Portraits of Path Finders and Path Creators.* New York: Palgrave Macmillan, 2015. DOI: 10.1057/9781137492470.0009.

▶

29

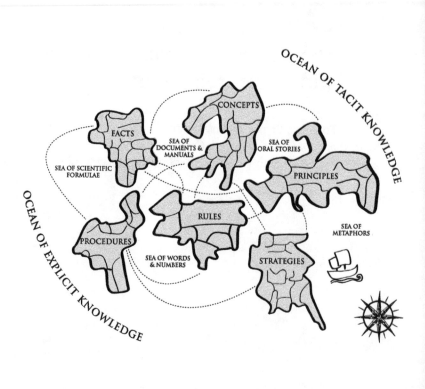

DOI: 10.1057/9781137492470.0009

At the very time when economists are debating if it really is the case that the innovation slowdown is an ongoing process, the embrace between science and entrepreneurship could be the real love affair in the first half of the twenty-first century. Both the path finders and the path creators have a part to play in that embrace. The former follow the fundamental rules for science; the latter ignore those rules and, if anything, adopt their opposites, showing themselves to be followers of the thoughts of Austrian-born philosopher Paul Feyerabend as expounded in his work Against Method (1975).

Early diagnosis of neurodegenerative diseases, battery-operated aircraft, objects that will have the energy to communicate with each other, teleportation of particles, the use of low-power light-emitting diodes to increase the yield of crops in greenhouses—we could add many more to this list—act as instigators of this union.

Lovers embrace that which is between them rather than each other.
(Kahlil Gibran)

Where do we stand with innovation?

Economic trends and the pace of innovation require not only reforms but also redefinitions of market conditions. Triggered by financial and real-estate bubbles, the Great Recession which started in 2008 has been a structural crisis, the solution of which demands structural transformations at micro (from person to household and neighbourhood), meso (from community to city and regions) and macro (from nation to international and global) levels. According to some economists, the weakening of innovation also requires structural adjustments.

Is there an ongoing slowdown of innovation as the economist Robert J. Gordon argues? In recent decades, have the myriad new products nothing to do with scientific advances? Is it just business ideas separate from science? (Phelps, 2013). The debate on these questions is still open. However, the fact remains that science and entrepreneurship were made for each other; their embrace can create the smouldering environment to create the love affair of the twenty-first century. As a consequence of the stronger ties between them, scientist–entrepreneurs and business–entrepreneurs driven by science grow in importance. This coincides with the abandonment of a return to pre-crisis growth trends of the world

DOI: 10.1057/9781137492470.0009

economy. The macroeconomic growth path is now unstable and it needs to be very different from that trodden even in the recent past.

Navigation across the Oceans of Explicit and Tacit Knowledge

In the two Oceans of Explicit and Tacit Knowledge, successive waves of intellectual resources, discoveries, inventions and technologies crash against the cliffs of the incumbent firms in the markets. While the erosive tidal action continues to take its toll, the interaction between science and entrepreneurship opens up unconventional entrepreneurial paths. Take, for instance, the case of 'Internet of Things'—also called the 'Industrial Internet', and 'Internet of Everything'—which involves trilateral collaboration between computer and social scientists from academia and industry, nascent entrepreneurs and business ecosystems—such as Intel, Cisco and IBM—in widening the range of connections between the physical world and virtual reality.

Science is a source of innovation. Entrepreneurs, the agents of knowledge-based innovation who convert that source into a resource (Drucker, 1986), share with the scientists traits such as creativity, questioning, eccentricity, a passion for unconventional ways to solve problems (the scientists); and seizing entrepreneurial opportunities (the entrepreneurs). In the pursuit of scientific discoveries and entrepreneurial opportunities, they depart from conventional pathways. It is the power of knowledge (first and foremost if absolute and coupled with precise measurement) that leads some of them, the path finders, to find paths that are not yet trodden. Others are driven by the power of ignorance, the equivalent of open-mindedness coupled with absolute uncertainty. That power gives them the strength to take a stance different from the traditional one and set out into unknown paths. They are path creators.

The encounter between the researcher and the entrepreneur in the scenario of 'Made in Italy'

Going back to its origins, 'Made in Italy' (see Chapter 9) unveils two facets. First, there is a set of entrepreneurial attitudes and motivations, which have built a bridge between the Renaissance—the major cultural and artistic movement that had its epicentre in the

DOI: 10.1057/9781137492470.0009

Florence of the Medici—and the Italian economic miracle that occurred between the end of the Second World War and the late 1960s. Second, there are the scientific specialties that sowed innovative seeds widely across the soil of entrepreneurship.

At the end of the Second World War, Giulio Natta, a professor in physical chemistry, met Pietro Giustiniani, a businessman who would become the leader of Montecatini, the largest Italian chemical company of the time. That meeting was in many ways the origin of a Nobel Prize and a blockbuster product. In 1963, the Nobel Prize for chemistry was awarded to Professor Giulio Natta for his discovery in the field of high polymers. Natta's research was taken up by industry and this led to the development and subsequent manufacture by Montecatini on an industrial scale in 1957 of a plastic material by the name of Moplen. Advertising accelerated its use and Moplen became a successful worldwide creation of 'Made in Italy'.

Unfortunately, with time the chain of 'Made in Italy'—from research, production, communication and consumption—has weakened due to poor maintenance and inadequate upgrading of relations between universities and industry. It is only in recent years that entrepreneurs and policy makers have become increasingly concerned with the management of the entire knowledge chain: from creation to diffusion, conversion and entrepreneurial exploitation of scientific and technological knowledge. Currently in Italy the knowledge chain has profound implications for higher education institutions and business schools which, to be successful, need to help companies create knowledge and be part of knowledge streams.

DOI: 10.1057/9781137492470.0009

4

The Unity of Opposites

Formica, Piero. *The Role of Creative Ignorance: Portraits of Path Finders and Path Creators.* New York: Palgrave Macmillan, 2015. DOI: 10.1057/9781137492470.0010.

▶

DOI: 10.1057/9781137492470.0010

The long-term action of a path creator that encompasses the short-term movement of a path finder, the latter to be found bordering a path creator: the appeal of the innovation lies in its conciliation of opposites.

Path finders are interested in what is closest to them. It is in this context of proximity that path finders gain new experiences and find new suppliers, co-producers and customers. The closer the proximity, the more the path finders tend to look alike and run along similar paths. Unfortunately, as the American science fiction author Frank Herbert wrote in his epic science fiction novel Dune (1965), 'The proximity of a desirable thing tempts one to overindulgence. On that path lies danger.'

The curiosity of the rambler who fixes his gaze beyond the visible horizon is the distinctive feature that characterizes the intellectual vigour of the path creator. Immersed in the domain of randomness and with the view of a quantum, the path creator experiences the uncertainty of being everywhere and nowhere, always and never. It will be the ability to master the uncertainty that will decide whether or not the path creator has the makings of a juggernaut.

We are sun and moon, dear friend; we are sea and land. It is not our purpose to become each other; it is to recognize each other, to learn to see the other and honour him for what he is: each the other's opposite and complement. (Hermann Hesse, Narcissus and Goldmund)

The duality of Heraclitus

In the universe of ideas, entrepreneurs orbit as planets around the Market Star with its magnetic fields created by actual and potential markets. As with Pluto, their orbits are irregular. They are either path finders or path creators. The furthest point of their vision is the horizon—the range and edge of one's knowledge. The horizon is not static, it moves, following a continuous series of changes in the positions and configurations of markets, cultural and social constructions in motion. Each and every player makes their path—that is, they decide whether to find a path to follow or create a new one. The aim of finding or creating a path is a challenge that explorers who want to build bridges to the future willingly accept.

Path finders and path creators do not remain fixed on one path leading to their goals. Whilst their variability in motion is different, the two groups of travellers are linked to a common destiny—the unity of

DOI: 10.1057/9781137492470.0010

opposites. The journey toward the creation of a path is the other side of the journey to find a path, and *vice versa*. Heraclitus (c 535–475 BC), a pre-Socratic Greek philosopher, called this duality the 'unity of opposites'. Their properties are opposite, but both are part of a whole which is in constant flux albeit not simultaneously.

The quest for stability through proximity

Path finders come to terms with the challenges of proximity. What they desire is to be close to customers who purchase and pay for their products and services. Although aware that their horizon is not static, their mission is to reach a state or period of stability during which no revolutionary change occurs.

Path finders are experienced pattern-finders who exploit the intelligible patterns of their deterministic domain of proximity, according to a table of rules that manipulates symbols in the same way as a Turing machine.[1] The framework of their research to find a path is 'Newtonian' economics, in which the mechanism of supply and demand operates, driven by individuals who act like automatons designed to maximize economic wealth. The trajectory of path finding is mechanically determined: the path finder misses the target only if the calculations are wrong; if there is a human error.

The quality or state of being proximate impinges on these calculations. Drawing on Boschma's seminal paper (Boschma, 2005), four types of proximity relations affect the path finder's assessment:

i.   Spatial Closeness (SC), which gives greater significance to stories that happen near to the agent;
ii. Social Proximity (SP), the source of attraction and trust;
iii. Organization Proximity (OP); and
iv. Cognitive Proximity (CP), the enablers, respectively, of common organizational practices and knowledge sharing.

Path finders take advantage of proximity relations to find new routes through which they can

a)   Improve efficiency and effectiveness— the 'optimization path', to be more efficient and cost-effective than competitors;

DOI: 10.1057/9781137492470.0010

b) differentiate— the 'specialization path', to be different from competitors; or

c) align— the 'neutralization path', to close the gap and be aligned with competitors.

The instability that creates new markets

The higher the degree of proximity, the more path finders read the same material and each ends up thinking the same way as others in the Unity of Proximity. The paths to be found are very similar to each other and not distant from the paths already travelled. Routine innovation prevails, taking place within the historical experience of the path finder. At the opposite end of the proximity spectrum, too little nearness makes the experience of the path finder more different than it otherwise would have been. However, in the search for a path conducive to entrepreneurship for new business ideas, the path finder is driven by negative values such as mistrust and opportunism. It is the negative values that divert the path finder in the search for an efficient and effective path.

In contrast, path creators navigate through the uncharted waters of randomness that are independent of accrued knowledge. What counts for them are the non-consumers who, by definition, cannot be counted. If path finders pursue the ultimate goal of a policy of stasis, the principal objective for the path creators is to achieve the instability that creates new markets. Path creators compete against non-consumption rather than seeking to attract existing consumers. In the field of path creation, they take a wide-ranging approach: from the 'forgotten man', who cannot afford to buy products and services already on the markets, to 'Mr Snob' who rejects anything that is not intended exclusively for the wealthy few.

With regard to the 'forgotten man', in post-Second World War Italy the Fiat 600 car helped substantially with creation of the path of growth that took the name of the 'economic miracle'. As the Glasgow Herald of 29 June 1959 declared, '[The] Fiat Six Hundred is the perfect combination of economy AND comfort AND performance, all in one attractively priced car! Small enough to banish parking troubles, but yet so generous in passenger and luggage space'. At the opposite end of the spectrum and also in Italy, Ada Masotti, an artisan skilled in the tradition of hand-crafted Italian corsetry, started the creation in her lingerie workshops in

DOI: 10.1057/9781137492470.0010

Bologna of a path that, 60 years later, would lead to the worldwide brand of luxury and feminine beauty: *La Perla*.

Think the unthinkable—the imagination is the fastest means of transport, impelling path creators to do things they do not understand; to go beyond the visible horizon. The mechanically determined process of path finding that typifies the proximity domain makes the brains of path finders more deterministic than they would like and gives way to randomness. In the randomness domain, devoid of antecedent causes or facts, there is no predictability, no certainty of individual outcomes. Each path creator is a depository of entrepreneurial energy transmitted in packets ('quanta') of indivisible ideas and used to create a path. The path creator is the player of a remarkable game, using an intuitive mind able to bring uncertainty to the level of probability. Because some outcomes are perceived to be more or less likely than others, the desirability of the action becomes stronger. The larger the number of path creators, the more likely it is that intuitive minds are at work. If, then, the randomness domain is sufficiently populated, it happens that there is a method in the creation of paths which, taken individually, leads to the conclusion that anything can happen. The behaviours of a large collection of random processes put in place by path creators become understandable.

Note

1   A Turing machine is a theoretical computing machine invented by Alan Turing (1937) to serve as an idealized model for mathematical calculation. See: http://mathworld.wolfram.com/TuringMachine.html

DOI: 10.1057/9781137492470.0010

5
Ambitious Entrepreneurs

Formica, Piero. *The Role of Creative Ignorance: Portraits of Pathfinders and Path Creators.* New York: Palgrave Macmillan, 2015. DOI: 10.1057/9781137492470.0011.

▶

Ambitions are the wings that enable entrepreneurs to fly. Amongst such individuals are the high-flyers with oversized ambitions and the largest possible wingspan for flying birds—comparable to that of a newly discovered extinct species, the Pelagornis sandersi. Why be so ambitious, then? Does it mean flying differently when, during flight, there are strong variations in the upward air currents and in the conditions on the ground below that is being watched? To behave differently would result in flight being regarded as 'right' and travelling on the ground as being 'wrong': by so doing, these high-flyers would be condemned to failure. This is what has been experienced by so many of those who boasted of their status as golden eagles in the sky of entrepreneurship.

Twenty years from now you will be more disappointed by the things that you didn't do than by the ones you did do. So throw off the bowlines. Sail away from the safe harbor. Catch the trade winds in your sails. Explore. Dream. Discover. (attributed to Mark Twain)

Pursuing opportunities beyond resources under control

Ambitions for growth are a compass that guides the entrepreneur in search of a path. In circumstances in which previous economic outcomes were resilient and self-reinforced current ones, highly motivated entrepreneurs do not break away from traditional paths. Exploitation of established practices and preferences continues to prevail over options of exploration. When circumstances change dramatically, the resilience and self-reinforcing effects that drove companies upward fade away and achievement of superior results depends thereafter on the pursuit of opportunities that lie beyond resources under current control (Eisemann, 2013). Ambitious entrepreneurs must move away from preferences, practices and sets of decisions in which traditionally they have been embedded. The search for alternative paths becomes an essential action.

The complacent entrepreneur

In path dependency mode, entrepreneurs derive advantage from what they already know. Exploitation involves playing the same game of

DOI: 10.1057/9781137492470.0011

incremental innovation, focusing intensely on the tasks that need to be mastered in order to optimize performance and efficiency. The underlying hypothesis is that consumers will be loyal and, therefore, will not abandon the brands they know and love. It is this complacency that blinds entrepreneurs to the facts. The complacent entrepreneur neither sees any need to create a new path nor recognizes threats from those outsiders who are generators of innovative pathways that reveal the 'next markets'. For example, in the world of mobile phones, one of the fastest moving industries of current times, complacency—epitomized by 'the smart phone cycle is alive and well'—and a lack of sufficient attention to threats (such as smart watches and internet-connected glasses)—has caused the derailment of two previously powerful players: BlackBerry (which in 2009 held about 22% of the smart phone market) and Nokia (in free fall, from its considerable market share of nearly 44% in the smart phone industry in 2008 to a meagre 3% in 2013; in its heyday in 2008 the Finnish giant sold some 468 million handsets, but sales dropped to about 120 million in the first half of 2013). Equally, filing for bankruptcy in 2012 was the punishment meted out to Kodak for having converted into the ignorance of the foolish the learned ignorance which, nearly 40 years previously, had led the company to the discovery of digital photography.

The deception of knowledge and the artfulness of ignorance

The iPad, by shrinking at an astonishing rate the space occupied by printed news, has brought about a drastic contraction of Finland's paper industry. Furthermore, the iPhone has knocked down heavyweight players such as Blackberry and Nokia.

You have reached 22% of the smart phone market, and your customers are loyal. Do you feel behind you the threatening breath of some alleged rival? No. Well, your name is BlackBerry and your madness is none other than the complacency syndrome, which has made you fall down the stairs. You are no longer one of the top five global manufacturers of smart phones and Nokia has accompanied you in your descent to the underworld.

Invoking the knowledge map—what you and your employees have learned and practised for years—is a useful exercise which, unfortunately, does not save you when the terrain changes dramatically. If you react by declaring that the ground, not the map, is

DOI: 10.1057/9781137492470.0011

wrong it means that you have entered the black hole of complacency syndrome. You have no way to escape.

A complacent feeling is the son of Great Knowledge, a mother who sees and provides. By believing in the knowledge map drawn for you, a new path was found to improve your business performance. Incremental innovation—that is, the ability to do even better what you already know how to do well—was your guiding light.

Seeking a new route to India and trusting the map—incorrect— made by Paolo dal Pozzo Toscanelli (1397–1482), it was by accident that Christopher Columbus discovered the New World. BlackBerry has not had the same good fortune. The 'new continent' iPhone remained unknown to the Canadian telecommunications company.

The stronger the assumption that the future will function much as today, the greater is the 'gravitational force' exerted by major technological advances (i.e., breakthroughs having strong technological effects), addressing under-developed market needs (disruption, with strong effects on markets) and market transformations (the game changer, with a strong effect on both technology and the markets). These 'strange' attractors, as envisioned by James Kalbach (2012) in his 'Four Zones of Innovation', make escape from the point of no return impossible. Current businesses slow down and never travel beyond the horizon: they are doomed to die. In contrast, visionary entrepreneurs are path generators who place the line of their horizon (i.e., the line beyond which their companies collapse) far from their perceived position.

And yet it moves

Path creation is a mode acting as a new source of knowledge, a deviation from the regular and usual course of the voyage, a singularity since there is no space (i.e., no market space) around it. Visionary entrepreneurs escape the restrictive success syndrome and explore uncharted waters beyond the horizons of existing technologies and business models. In the process of exploration they are guided by creative ignorance; or to state it slightly differently, they are keen to uncover facts and data that disprove the assumptions resulting from acquired knowledge. As Enrico Fermi argued, if you do an experiment and 'the result is contrary to the hypothesis, then you've made a discovery'. Those who create paths do not therefore invest in a strong hypothesis which requires them to

DOI: 10.1057/9781137492470.0011

carry out tests likely to prove it. Experimental capital and the availability and quality of experimental labs, where questioning is not restricted by abstract ideas accepted as simple and irrefutable facts, are preconditions for propelling the process forward and gaining rapid feedback (Curley and Formica, 2013). We must not forget that *experiri* (to try) was the key with which Galileo Galilei's alleged declaration of *Eppur si muove* ('and yet it moves'), regarding the movement of the Earth around the Sun, opened the door to the future of science.

DOI: 10.1057/9781137492470.0011

6

Creation Out of Nothing

Formica, Piero. *The Role of Creative Ignorance: Portraits of Path Finders and Path Creators*. New York: Palgrave Macmillan, 2015. DOI: 10.1057/9781137492470.0012.

▶

DOI: 10.1057/9781137492470.0012

WHITE IN
PROGRESS

DOI: 10.1057/9781137492470.0012

> *How do we create something from nothing? Does what we do come from opening at random the book of knowledge and then being forgetful enough not to be stifled by the knowledge gained? This may well be the case; but the first step toward path creation is awareness. We do not allow ourselves to be led by knowledge, which is experiential and stored in our head, but by awareness, which is intellectual and in our mind. Being fully aware and relying on our thoughts and intentions, we can recognize and address the challenge of the secret and narrow path that is difficult to traverse—the 'Atrapos' of the ancient Greeks—and which leads us to bring previously unknown ideas into existence.*
>
> 'Survivorship bias is a profound insight into how we confuse the known unknowns and the unknown unknowns.' (OfTwoMinds Blog, The Unknown Unknowns And Survivor Bias, 17 June 2013)

Venturing down the *atrapos*

'Atrapos' is the word used by the ancient Greeks to describe a secret and narrow passage that connects two worlds otherwise isolated from each other. On the one hand there is the known world, depicted in the knowledge map; on the other hand there is the world of nothingness—what our knowledge defines as empty space or void. Who ventures down this 'atrapos'? We can say that characters such as Samuel Beckett's Estragon and Vladimir, protagonists of *Waiting for Godot*, do not tread such a path. They would like to, but.... As Estragon says: *Well, shall we go?* And Vladimir responds: *Yes, let's go.* Yet: *they do not move.* The two separate worlds remain unconnected.

Knowledge acquired through past experience is of special interest when seeking a path upon which to travel rapidly. The reality that people perceive is that of the winners whose performances are well documented. The winners deliver unimaginative results; the better they are, the harder it is to move to the mental space of not knowing, where creative results are pre-eminent. Brilliant, albeit boring, results increase the fear of trying to tackle the territory of ignorance. Therefore, path finders are at a disadvantage compared to path creators who benefit from the fact that people dismiss as genuine the more demanding but much more enthralling path, deny that it is relevant or are reluctant to accept its relevance. What is more, information on the creators who have perished along the

DOI: 10.1057/9781137492470.0012

path is mostly non-existent. The losers vanish from the records. This bias is a near insurmountable barrier to knowing what we don't know.

Tambora syndrome

Rejection of the new path is a force that acts as a brake on the spread of the news. Those who have fallen into the trap of the knowledge of the market in which they are embedded are subject to what we might call 'Tambora syndrome'—derived from the name of the Indonesian volcano whose catastrophic explosion in 1815 was reported in a small news item in *The Times* in London seven months after the event, despite the fact that by that time 'Tambora's effects were already being felt' (Bryson, 2003). By the same token, incumbents experience the potential of high-technology progress (breakthrough) and high level of market impact (disruptive) innovations contained in the creation of a new path only after a considerable delay. They cannot sense a threat on their visible horizon. As a consequence, start-ups that follow the new path have a competitive edge over established companies holding dominant market positions. Little wonder, then, that Tambora syndrome has claimed countless victims, of which the following are but two of the many examples from the past 30 years of such striking effect that they caught the attention of the media. First, Wang Laboratories, which remained committed to the word processor when IBM launched its first personal computer; and, second, BlackBerry which considered the iPhone that Apple introduced in 2007 to be no more than a simple toy (Nocera, 2013). Blackberry did little more than experience again the error of perspective that Western Union had made more than a century before in defining Bell's telephone as an 'electrical toy'.

Game changers

Figure 6.1 shows the two sets of characteristics associated with path finding and path creation. In the case of path finders, the magnitude of 'I don't know' falls within the domain of their knowledge: it is 'known ignorance'. Path finders are guided by the experts' structured thinking of how to play the game they know, who find the path using eyeglasses of wisdom, experience, the preferences from past experiences, and careful

DOI: 10.1057/9781137492470.0012

planning. For path creators, the magnitude of 'I don't know' is unconstrained by knowledge. They operate outside the comfort zone of 'known unknowns', moved by the freedom of not-knowing and receive support from the unlikely quarter of non-experts. These latter help in deciding how to navigate unpredictable situations by wearing the eyeglasses of naïveté, creativity out of nothing, open-mindedness, emotion and fearless conviction, leading to the discovery of surprising paths toward as-yet non-existent markets.

Path creators change the game

1. Neil MacGregor's case

When Neil MacGregor, the director of the British Museum, was appointed director in 2002, it was reported that 'Many worried whether [he] had the administrative skills to turn things round; his reputation was primarily that of a scholar rather than a manager. In fact, he convinced the museum's trustees to let him shape the job to suit his strengths. This meant allowing him to become a "cultural diplomat", bringing exhibits from far-flung parts of the world to the museum and delegating administrative chores to others. Mr MacGregor became not just a successful director of the museum but also an unusually powerful one.'

Source: Philip Delves Broughton, 'The path to power and how to use it,' *Financial Times*, July 16, 2013.

2. 'Lower your radio please'

Path finders are attached to the consolidation and refinement of knowledge, not revolution. In short, they use their instruments to detect consumers in existing categories of products and services within the horizon defined by their knowledge map. Path creators, meanwhile, chart paths where they do not meet travelers (i.e., consumers): their instruments are thoroughly non-intuitive, designed to detect technologies and business models which make and reveal customers beyond the visible horizon. They operate out towards the true horizon, a great deal further than path finders.

The radio ushered in a new era of communications. Guglielmo Marconi's work with radio waves added the cognitive dimension of mass communication to Henry Ford's tangible dimension of vehicles

DOI: 10.1057/9781137492470.0012

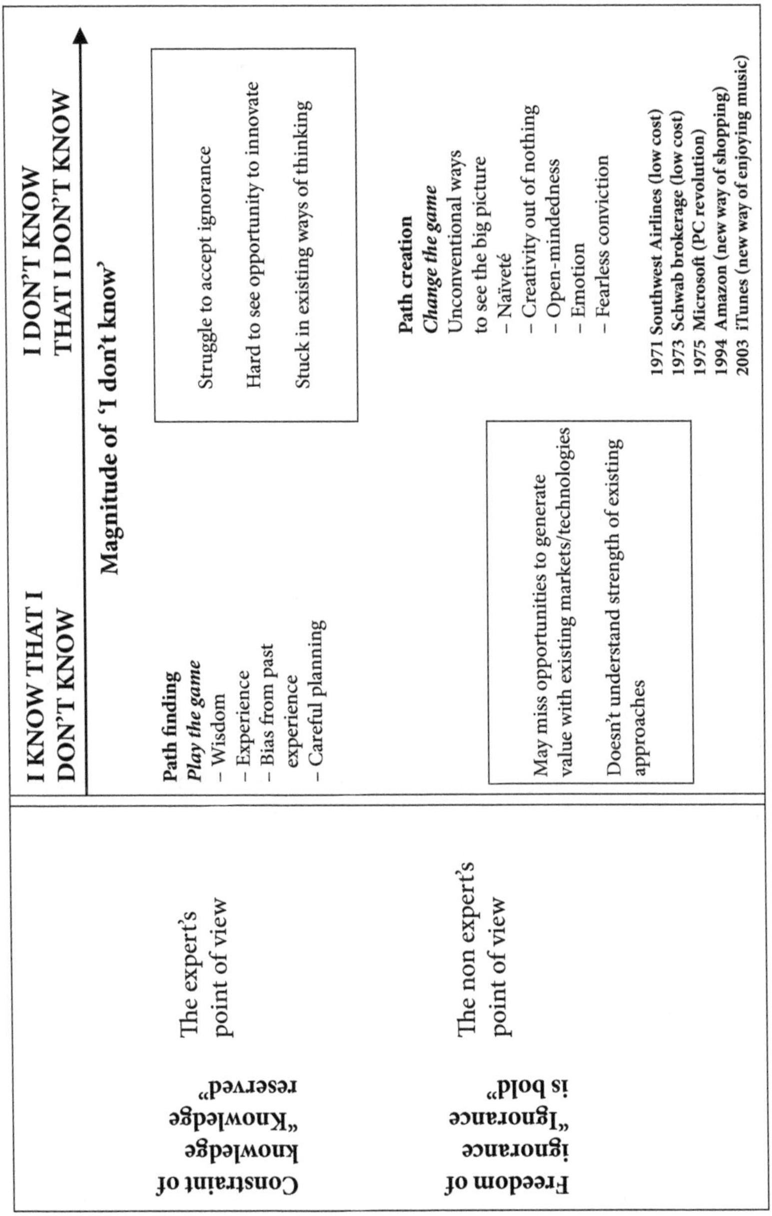

FIGURE 6.1 *Path finding and path creation: two sets of characteristics.*

and roads. As we wrote in our *Stories of Innovation* (Formica, 2013), Marconi was the protagonist of the interlude between the Fordian modern age of mass production and the view the Marconian waves had gradually revealed of the unprecedented social and economic landscape of the twenty-first century.

In Fascist Italy, owning a radio was a status symbol. Listening to it at full volume was used to make neighbors who did not have a radio envious. As a result, the song 'Lower your radio please' was a big hit of the time. However, it was not always possible to ask your neighbor to turn down the volume. At bathing establishments—increasingly popular in the 1930s and favored by Benito Mussolini—using the radio loudspeakers was the only way to listen on the beach. Take the radio from home? Impossible. Maybe someone was dreaming of it: but there is no demand for what does not exist.

You must wait for a path creator, for whom framing questions (in our case, 'what if I design a device that people could carry with them to hear sounds and voices?') is more important than finding and offering answers. The new path became visible to everyone in 1960 when Sony introduced its first transistorized radio, small enough to fit in a jacket pocket and able to be powered by a small battery. Thus was started the age of mass customization of sounds and voices.

Before the question 'What if I …' was asked there was no market space. The only space that existed was shaped as the creators of the Sony's transistorized radio proceeded along the path that they traced, proceeding step by step. As Akio Morita, co-founder of Sony, argued in February 1979 at Sony Headquarters about the Walkman years later, 'This is the product that will satisfy those young people who want to listen to music all day. They'll take it everywhere with them, and they won't care about record functions. If we put a play-back-only headphone stereo like this on the market, it'll be a hit.' Sony's achievement looms large on the path finders' horizon. Many of the path finders had limited themselves to making improvements to the 'traditional' radio. 'Lower your radio please' lost its original meaning.

At the start and for an indefinite time span (because of unanticipated threats that make the time signature different from path to path), the new path is very short and narrow when compared with existing ones.

DOI: 10.1057/9781137492470.0012

However, since opening this new path leads to the propagation of a new venture that has been disclosed, and the resulting ramifications, the potential spaciousness of the new path is orders of magnitude greater. Among the incoming entrepreneurs there are those who *ex post* (i) recognize that the idea behind the path creation was not yet on their radar or (ii) were unaware of knowing it ('unknown knowns', things we don't know that we know, a term attributed to Slavoj Žižek, 2006) or (iii) deemed the idea not ready because they had not realized its commercial value.

DOI: 10.1057/9781137492470.0012

Part II
The Portraits

▶

DOI: 10.1057/9781137492470.0014

7
Experimental Theatre: Path Finders and Path Creators Take to the Stage

Formica, Piero. *The Role of Creative Ignorance: Portraits of Path Finders and Path Creators.* New York: Palgrave Macmillan, 2015. DOI: 10.1057/9781137492470.0015.

Trying something new leads path finders and path creators take to the stage in the experimental theatre, a laboratory where experiment-based approaches to path finding and path creations are explored: from the creation of ideas and then continuing along a route with two significant milestones in the process of sharing and refining new ideas. In the backdrop of the narrative that is played out, the strait whose stormy waters flow next to the rocks of the 'Scylla' of running a calculated risk and the 'Charybdis' of uncertainty about the entrepreneurial endeavours comes into view.

It is the highly radical and innovative dramatists—Brecht, Ionesco and Beckett for instance—who ultimately have the greatest impact. Their brave and pioneering experiments have changed profoundly the way of doing theatre. Similarly, path creators are the playwrights and comedians who break conventional habits and stereotypes in the theatre of experimentation for innovative pathways.

Innovations do not arrive fully-fledged but are nurtured through an experimentation process that takes place in laboratories and development organizations. (Stefan H. Thomke, 2003)

On the stage of the Experimental Theatre

Path finders and path creators are actors who take to the stage of the Experimental Theatre, where they try something new. Acting as experimenters, our performers start by testing beliefs and ideas that they want to turn into a business. By running experiments, business ideas move from an embryonic and rudimentary state to full manifestation in the form of new ventures. In particular, conducting experiments provides opportunities to learn how to mobilize their new ideas to anticipate change, take a chance on the future and structure their business in such a way that it can successfully gain access to the marketplace.

Entrepreneurial experimentation is the theatrical method that relates a business concept to an experiment. The idea is that the concept originator is then stimulated to build on the concept in the light of experimentation. Since the opportunity cost of experimentation decreases with the increase in value of a new business venture, high-performing players,

DOI: 10.1057/9781137492470.0015

whose high value-added business propositions bear low opportunity costs, are likely to show greater propensity to conduct experiments than are actors who are pursuing low value-added activities.

Path creators emulate Brecht, Ionesco and Beckett whose writings overturned previous conventions and created new kinds of theatre. The more daring the experiment conducted by them, the greater the chance of a path-breaking result.

Experimentation in three acts and...

'Idea building' is the first in the series of experimentation. This burst of inspiration has the advantage of creating a language that moves the experiment forward, thanks to the formulation of a strategy and the interaction with other actors. At the end of this stage, a prototype becomes available.

'Idea reformulation or re-evaluation' is the second set of experiments. Experimenters receive feedback from a small number of potential customers, following which the original business concept, with its assumptions, is reformulated or re-evaluated.

'High growth potential' is featured in the third stage of experimentation, which the experimenters manage with the aim of building a bridge between the very small group of initial customers and the wider platform of less adventurous, more practical buyers.

Two scenes: analogical and conjectural modes of experimentation

Path finders and path creators are, respectively, exposed to analogical and conjectural modes of experimentation. The process of finding an acceptable path is generated by analogy with paths previously trodden or observed from which the finder has received positive feedback.

The analogical mode accomplishes the task of shrinking the area of known ignorance about the finder's idea. Analogy-based reasoning can be applied to the known unknowns of a new concept whose business domain has attributes in common with those of another domain. The analogical method is represented by the typical case-based approach: past cases from a different domain are used to highlight possible solutions for problems incurred with the new idea.

DOI: 10.1057/9781137492470.0015

The conjectural mode proceeds by trial (involving the spontaneous, serendipitous discovery of building blocks for the business idea under experimental scrutiny) and error (leading to the elimination of elements that are demonstrated to be inappropriate). High-expectation business concepts sail into uncharted waters, exhibiting unfamiliar traits of novelty and complexity; purposeful ignorance exposes path creators to a voyage into unknown unknowns. High-expectation propositions, therefore, cannot be investigated with the mental apparatus of analogical reasoning and, specifically, case-based reasoning. When no apparent rules or commonalties can be applied, trial and error is the approach that can support an imaginative and conjecture-based process of discovery. The major cost of this approach is the time invested in arriving at a solution from the iterative process, triggered by selecting what *ex ante* looks like the most suitable choice. If something does not work, the process has to be iterated until an appropriate and acceptable answer is found.

DOI: 10.1057/9781137492470.0015

8
Portrait of Path Finders

Formica, Piero. *The Role of Creative Ignorance: Portraits of Path Finders and Path Creators*. New York: Palgrave Macmillan, 2015. DOI: 10.1057/9781137492470.0016.

▶

DOI: 10.1057/9781137492470.0016

> *Where there is knowledge, the light is bright. What really matters to path finders is the talent for absorbing knowledge. The thought that the intense light can bring with it deep shadows does not occur to them.*
>
> *The knowledge map is the compass that guides the path finders' footsteps. Each map is a representation of knowledge assets, artefacts, people who hold those assets or are the creators of artefacts. It also provides an overview of the patterns of knowledge flow.*
>
> *Making use of one or more maps, the path finders experience how to find a path through studies and analysis, performing surveys, and involving individuals or groups of people in action research, all designed to achieve the result they expect.*
>
> *The expert who has a stubborn belief in an idea and therefore digs into the well of knowledge, having a high level of skill in a certain subject, stands in the full height and depth of cognition.*
>
> *Where there is much light, the shadow is deep.* (Johann Wolfgang von Goethe)

The best of all possible worlds

Path finders look for a sense of security, which matters to them both as an ideal of life and as an asset. They combine innovation with words such a continuity, conservative, incrementalism and viability. Nassin Taleb (2012) has invented the word 'fragilista' to describe a person who loves 'order and predictability', and who suffers as a result of 'random events, unpredictable shocks, stressors, and volatility'. Therefore, in accordance with Taleb, the sense of security poses the risk to path finders of facing severe side effects which weigh heavily against the small benefits arising from their actions. We might say that what comes to inspire path finders is the 'Golden Age of Security', admirably portrayed by Stefan Zweig in his autobiography *Die Welt von Gestern* (1944) (Zweig, 2009). As Zweig said, age is 'an infallible road to the best of all possible worlds'. Security leads to an overestimation of the power of the knowledge.

Finding a path is an exercise in exploitation that requires knowledge. Path finders are guided by principles and rules established and codified in their knowledge maps. With this lens, they look at the visible horizon—the line at which the 'earth' of today's businesses appears to meet

DOI: 10.1057/9781137492470.0016

the 'sky' of the businesses-to-be in the making. The path finder is an individual or a collective 'knower' who has a business idea together with a depth of knowledge of its field of application. The 'knower' closes the gap between known but unanswered questions about that field and the answers to those questions. Both questions and answers are influenced by information from survivors—primarily the winners—on the area of business under investigation (Freakonomics, 2009).

The Ptolemaic 'knowledgists'

Innovators, as inhabitants of Today's Island, surrounded by the seas of uncertainty and risk, look to the future through different sets of lenses. Many amongst them are path finders—Ptolemaic 'knowledgists' whose movements are determined by the knowledge map, an immutable reference system. Everything revolves around it. They go in search of paths within the boundaries of the map, setting in motion processes of discovery or fact finding by study and experience (Figure 8.1).

In the same way as the great generals like Alexander the Great, path finders try hard to maintain high levels of all the attributes both created and demanded by knowledge acquired through education and practice. Path finders—commercial innovators, staunch supporters of the claim that existing knowledge is the source from which most innovation stems—try to avoid repeating past mistakes and analyze these experiences as a means of shaping the future. Doubt puts them into a state of despair in relation to the knowledge mastered and leads them to believe that they are not competent to respond to the questions posed by knowledge. This state of mind encourages path finders to extend their knowledge maps toward adjacent branches. Through this process they create effective combinations in order to find new paths.

Path finders pay special attention to the knowledge maps thickly dotted with statistical data of both the recent and remote past. They make use of lenses that show future scenarios consistent with the trends of past data. Thickened by the data, those lenses now seem even more reliable to the innovators who wear them, able to extend their ability to predict. The distant future appears to be closer. However, if placed on routes that albeit new are no more than the same, less-risky incremental variety, path finders miss the future once changes become revolutionary rather than evolutionary. They become vulnerable to path creators—the

DOI: 10.1057/9781137492470.0016

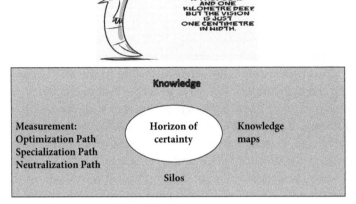

FIGURE 8.1   *The Ptolemaic 'knowledgist'*

new challengers who threaten the successful strategies, business models or products of path finders.

Reach Asia by sailing west

Those who want to find a path believe that, by holding knowledge maps drawn by the scholars of their time, it is possible to detect the future course of events. Once those scholars have gained mastery in one or more disciplines, maps drawn by or through them become known to the path finders. Henceforth they firmly believe that the best and most attractive ideas on how markets work are embodied in the knowledge maps at their disposal. The existence of a trend identified by scholarly expertise directs the search for a path within the territory circumscribed by the 'horizon of certainty' (Figure 8.1).

In his *Meditationes de prima philosophiae* published in 1641, René Descartes (1596–1650) wrote that 'Dubium sapientiae initium' ('Doubt is the origin of wisdom'). However, a body of ideas about a particular subject or discipline can become commonplace knowledge and an accepted standard. Thus conventional wisdom gains the upper hand and doubt fades away. What is now visible is the 'horizon of certainty', despite the fact that the reality is the 'horizon of doubt'. Risk finance,

DOI: 10.1057/9781137492470.0016

bank credit, extensive and close relations with the incumbent players, and consolidated knowledge all propel the process of exploration.

Once possessed by an idea, which has perhaps suddenly come to mind, or which was there waiting passively for the right moment to arrive, we can apply our knowledge to it or we can exploit our ignorance. Possessed by the transformative idea to find a westward route to India, Christopher Columbus acted according to the knowledge of his time. Of the many innovation paths available, Columbus opted for that governed by a set of characteristics in accordance with the current thinkers of his era. It was Paolo dal Pozzo Toscanelli (1397–1482), an Italian mathematician, astronomer and cosmographer who exerted a decisive influence on the decision by Columbus to search for the East Indies by sailing west. 'Toscanelli's map [of the World] had miscalculated the size of the earth which resulted in Columbus not realizing initially he had found a new continent' (*Wikipedia*, entry on Paolo dal Pozzo Toscanelli, http://en.wikipedia.org/wiki/Paolo_dal_Pozzo_Toscanelli). The book *A Literary and Historical Atlas of America* by John George Bartholomew (1911) presents a hypothetical reconstruction of the Lost World Map of 1474 produced by Toscanelli, which clearly shows the position of the American continent, which he ignored:

ATLANTIC OCEAN, TOSCANELLI, 1474

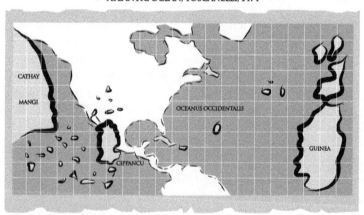

DOI: 10.1057/9781137492470.0016

9
Path Finders of 'Made in Italy'

Formica, Piero. *The Role of Creative Ignorance: Portraits of Path Finders and Path Creators.* New York: Palgrave Macmillan, 2015. DOI: 10.1057/9781137492470.0017.

▶

DOI: 10.1057/9781137492470.0017

'Made in Italy' is a substantial artefact comparable in its mission to that of the Lego company, whose chief executive, Jørgen Vig Knudstorp, said that Danish toy maker would never become a fully digital company: 'The one thing is that we never leave the physical brick. Our standpoint is that physical play is extremely important. Then I see digital as an extra experience layer'.

The label 'Made in Italy' stirs emotions and transmits the cultural values of its many territories. As research at Bocconi University showed, the fashionable, elegant and outstanding companies of 'Made in Italy' are the bearers of emotion that have withstood the waves of the Great Recession and even ride them successfully. But beauty is not enough if Italy amongst the major European countries has lost a quarter of its manufacturing output since the crisis began in late 2007.

Digital innovations must be the handmaiden to beauty, with 3-D printing, e-commerce platforms and applications in fashion, furniture, agro-food, mechatronics and many other facets of 'Made in Italy' hybridizing the physical world with the virtual one.

> 'Cheshire Puss', Alice asked, 'Would you tell me, please, which way I ought to go from here?'
>
> 'That depends a good deal on where you want to get to', said the Cat.
>
> 'I don't much care where...' said Alice.
>
> 'Then it doesn't matter which way you go', said the Cat.
>
> '...so long as I get somewhere', Alice added as an explanation.
>
> 'Oh, you're sure to do that', said the Cat, 'if you only walk long enough'.
>
> – (Lewis Carroll, Alice in Wonderland)

Where 'Made in Italy' wants to go...

Alice, walking through a wood, cannot decide where to go. At a crossroads of production and innovation, path finders of 'Made in Italy' know what to do: they want to seize the good fortune that results from harnessing the purchasing power of 2–3 billion new consumers in the world. The road to take is that of production, less difficult to travel if they were willing to try to increase the pace by opting for incremental innovation. Theirs is a search for convergence between production and innovation to be achieved by reading with care the knowledge maps they

DOI: 10.1057/9781137492470.0017

possess. Like the Germans, the path finders of 'Made in Italy' rely on making products of increasing quality for which consumers are willing to pay premium prices. Within their visual horizon one can glimpse e-commerce, hitherto—as the numbers provided by the financial daily *Il Sole 24 Ore* emphasize—the exclusive preserve of a few players only. The edition of that newspaper published on 27 August 2014 included an article by Giuliano Conti headed 'If it were a brand "Made in Italy" would be the third in the world', in which Conti states that:

> The searches conducted on Google in the first half of 2013 show that 'Made in Italy' and its key sectors grew by 8% compared to the same period of 2012, with peaks in Japan (+29%), Russia (+13%) and India (+20%). Despite these numbers, only 34% of Italian SMEs are online with their own websites; only 4% of Italian companies with more than 10 employees sell at least 1% online, compared with 12% of French and Spanish and 14% of UK and 21% of German companies; the best 20 Italian companies operating online cover together 70% of the turnover of the Italian e-commerce: the first 50, 86%.

…and where it might wish to go if not hampered by the burden of the guilds and the limit of meritocracy

'Made in Italy', replete with cultural relevance, nurtures sentiments of grace and beauty, variety and abundance, and leisure. However, 'Made in Italy' is not yet at the point of advocating the idyllic society for which John Maynard Keynes yearned for the grandchildren of his generation. The British economist argued, in his essay 'Economic possibilities for our grandchildren' (1930), that the obstacle was 'The love for money as a possession [which is a] mental disease.' In accordance with Keynes' thoughts, infinite spaces open up for path breakers to revise the moral code so that 'the love of money [becomes] a means to the enjoyments and realities of life'.

Unbridled love for possession of money has not spared the many players, with those of later generations at the forefront, of 'Made in Italy'. Corporatism, which reigns supreme in contemporary Italian society, promotes and encourages this attitude. As Edmund Phelps, Nobel Prize in economics, reported in an article in the Financial Times (Phelps, 2014a), the fixation on becoming rich is due to the values of corporatism: in the case of Italy, primarily materialism fuelled by supported interest groups. The perverse combination of 'pork-barrel' contracts and short-term results implemented by these groups stifles innovation, and

DOI: 10.1057/9781137492470.0017

particularly the innovation that would arise from path creators. Their ground-breaking innovations would shake the foundations on which corporate interests are based.

It is difficult to identify a nation where, more so than in Italy, thinking framed in the knowledge of professional practices dominates with regard to the decisions and choices that shape society and the economy, forcing both into a perpetual *status quo*. It could not be otherwise, given the burdensome legacy that Italy has inherited from the city-states—perhaps the best known of which is Florence—which flourished in the country between the ninth and fifteenth centuries. It was in those cities that the guilds of artisans and merchants were prominent in undertaking the task of setting and monitoring the knowledge and practices of their trades. The guilds are now Trade Associations and Professional Bodies whose structured thinking is the outcome of knowledge gained in the professional schools of the universities that educate future professionals to take up career opportunities in the restricted world of interest groups.

There are conflicting opinions among historians on the economic consequences caused by the guilds. One school of thought attributes a positive impact on the economy to the guilds, far outweighing the benefits which were enjoyed by their members. In contrast, another school asserts that the guilds were responsible for privileges and the enjoyment of special positions that favoured their members, and that the conservatism of the guilds has been a restraining force on innovation. Groups of insiders, who have invested heavily in maintaining the status quo and therefore resist the pressure of the reforms, stifle the creativity of 'Made in Italy', as illustrated by the following data. This is especially the case for those sources of creativity existing at the confluence of digital technologies and business and organizational models.

Italy: ranking in the Global Innovation Index (of 143 countries)	31
of which	
Creative Outputs is ranked at	48
of which	
Intangible assets is ranked at	113
of which	
ICT and business model creation is ranked at	108
and	
ICT and organizational model creation is ranked at	116

Source: The Global Innovation Index 2014—The Human Factor in Innovation.

DOI: 10.1057/9781137492470.0017

In Italy today an excess of the 'rentier economy' and a deficit of innovation are the two evils transmitted as if by contagion, caused by the guilds in their various forms, to the entire economic body.

Even worse is that a third handicap even more inconceivable has arisen: gerontocracy. Because the strength and influence of thinking framed in knowledge practices are considered to be positively correlated to the age of the thinker, gerontocracy and not meritocracy rules the country. A recent Gallup poll revealed the faint imprint of meritocracy in Italian society. While 54% of Italians believed in a meritocratic society, it was 69% in France, 74% in Germany, 78% in the UK and 89% in the United States. Looking to the East, the gap between Italy and other nations is even wider, with China showing 93%, India 90% and Australia 82%.

Moreover, meritocrats themselves, when in positions of responsibility, are so blinded by the charm of being the holders of knowledge that they are disinclined to be associated with creative ignorance. So, if the guilds stifle innovation *tout court*, meritocracy does not allow those who have an interest in creative ignorance to drive a wedge between the meritocrats. There is nothing the creative ignorant can do other than try to move in search of lands of opportunity—in the direction of the nations of Northern Europe, North America and the vast Asia-Pacific region.

The combination of these conditions will slow down and even stop the flow of venture capital into the metaphorical fuel tank of innovative talents, whether path finders or path creators. Thus impoverishment of innovative business initiatives coexists with an abundance of wealth held by Italian households. Italy represents 1% of the world population and accounts for 3% of global GDP, but its share of the family wealth is much higher, equal to 5.7%—on average, €350,000 per household. Furthermore, with 10% of households owning 45% of the total wealth, the group of individual investors could be significantly large: it is not, and the country remains in 'oxygen debt' of innovative entrepreneurship. This bodes ill not just for would-be path finders and path creators, but also for the nation.

DOI: 10.1057/9781137492470.0017

The Italian hornet and the lame duck

DOI: 10.1057/9781137492470.0017

What folly! The bumblebee flies without knowing why and seemingly contrary to the laws of aerodynamics. Not long ago, a rare species of bumblebee, from Italy, was revealed in the 'Big Store', the Italian version of the Marx Brothers comedy film of the very early 1940s. After World War II, a bee named Italy was flying in the skies of the Western economy. In spite of its aerodynamics, this particular insect was able to gain altitude, travel fast and stay on course thanks to wings provided by craftsmen and small entrepreneurs who inaugurated the era of 'Made in Italy'. At the turn of the twentieth century, the double-barrelled shotgun of innovation and bureaucracy coupled with political corruption hit the bee and, with it, many 'animal spirits' with their entrepreneurial vision, once so vital, have also dropped from the skies.

The Italian economy is what many consider to be a 'lame duck'. 'Made in Italy', the strong leg of the economy, still enjoys the same good reputation it always has, even when times are tough—as is the case in the current international economic climate. However, Italy's standing with regard to globally integrated enterprises is much weaker: from multinationals converted into entrepreneurial ecosystems to start-ups, Italy is losing ground in a world changing all too quickly in order to keep pace with the digital revolution.

Lessons from Ireland

Ireland is another country which has posed problems for the Eurozone. Nevertheless, the performance of the Irish economy can teach Italians something about how to strengthen their position in the field of technology.

In 1990 Fruit of the Loom first established its manufacturing base in Ireland for the production of T-shirts. Twenty years later, an orchestra of global companies has formed around the Celtic harp, many of which have already established R&D laboratories. These companies include Intel, Pfizer, Merck, Google, Apple, Siemens, IBM and Microsoft, together with the most recent wave of social network businesses such as Facebook, LinkedIn and Twitter. Ireland is currently home to nine of the ten largest global technology companies and, per capita, attracts foreign direct investment ten times greater than Italy.

The Italian ruling class has embraced the pessimism of the critics of multinationals—that such companies are too capital-intensive, do not

DOI: 10.1057/9781137492470.0017

create enough employment, are ready to flee at the first sign of unprofitability, and are not such an effective vehicle for export. Unfortunately, as a result, the Italians have lost sight of what some of these multinationals have become. Beautiful butterflies eventually emerge from ugly caterpillars and, similarly, multinationals that become global enterprises can nourish and give rise to new generations of entrepreneurs. Ireland has three main attributes gathered over time: a tax system generous to global enterprises; a relatively young workforce, adaptable, educated and with marketable technological skills; and a bureaucracy which helps foreign firms and facilitates start-ups. Despite the strong turbulence the Irish economy has faced in the last five years, including a net debt-to-GDP ratio that has jumped a massive 93 percentage points, the result is that it is possible in Dublin now to invest in nascent entrepreneurs, generate employment and raise the companies and individuals concerned on the world's new ways of conducting business. To offer but two examples:

▸ New investment by Intel of US$500 million for a new plant for the production of computer chips; and
▸ The decision by Twitter to establish its international headquarters in Dublin.

The 'Made in Ireland' brand is alive and well in these industries.

Italy is a 'Little Ireland' with regard to these matters. However, the future will not revolve around a 'Made in [a country]'. It will instead be a world without boundaries that will ultimately lead to a 'Made in the World' brand. Global business champions are seizing excellence wherever it is available and weaving networks of international talent. They are becoming worldwide integrated business communities. As the number of customers increases, especially among the new wealthy classes—from China to India, from Turkey to Brazil—the 'Made in' gives way to this symbolic 'Made in the World', made up of many 'in' centres of excellence (from design to logistics). Countries must act now to ensure they can secure a strong foothold in this new business environment.

To do business and start new businesses in Italy one is confronted with a very hostile terrain compared to other countries. The World Bank's rankings on the ease of developing a business and on business start-ups place Ireland in 10th and 13th place, respectively. Italy, by the same measures, ranks 77th and 87th, respectively. When compared to other international locations, international companies fail to give even a second thought to Italy as a possible base. In contrast, Dublin and

DOI: 10.1057/9781137492470.0017

other locations are regarded as crossroads of international talent. Italy lacks catalysts such as the Enterprise Ireland agency—capable of attracting tens of millions of Euros to be allocated to nascent entrepreneurs. Italy does not place the same value on the creative class (professionals of the knowledge economy) which, according to estimates by Richard Florida, represents 13%, versus 26% in Ireland, of total employment. Italy is lagging in the provision and take-up of higher education and possession of higher education qualifications. In the 24–65 age group, for instance, 54% in Italy have a secondary school diploma, compared to 72% in Ireland. The gap remains significant even when the age group is reduced to 25–34 years: 70% in Italy and 86% in Ireland. Italy lags behind in the number of degrees awarded covering mathematics, science and computer science, qualifications that, according to the Eurostat Yearbook for 2010, saw Irish youth aged 20–29 as the third best in the European Union. The final, but perhaps most critical factor, because it relates to the attitudes of the younger generation that will shape the future of the nation, is that Italians cannot liberate themselves from the safety of the state-guaranteed value of qualifications awarded by universities. This legal value, as Luigi Einaudi, who served as the second president of the Italian Republic, had noted some 50 years ago, 'is a concept completely foreign to the Anglo-Saxon countries'.

So it is that in Ireland the students aspire to form the new entrepreneurial class. An example of this is the recently established Trinity Economic Forum, which will help shape the future direction of national economic policy. In contrast, Italy remains educationally divided, with university students protesting against whatever the Minister for Education of the day implements as they go through education suffering the anguish of intellectual unemployment.

Italy cannot transplant the traits of the Irish economy but, with persistent effort, it can learn from the Irish and once again embrace the entrepreneurial spirit.

'Made in Italy': what does it really mean?

In the 1950s, a time of pioneers of trade without borders, this was a disturbing and perhaps less than obvious question. The protagonists of that time were first-generation entrepreneurs of small businesses who answered the question by initiating a process of assimilation with the

DOI: 10.1057/9781137492470.0017

economic achievements of the Free World. 'Made in Germany' was their benchmark—a perfect example of how much an economy devastated by war, such as that of Germany, could rise again in the Free World. The work of these first-generation entrepreneurs succeeded because assimilation did not mean passive levelling down to the values of others, but rather the combination of those values with the emerging culture of the small and flexible company.

Rather than being lifted into the air by German giants, the Italian dwarves clambered over their shoulders to see more and, having more acute vision, even further. While remaining proud of their performance, the special attributes and the positive aspects of diversity of family businesses consisting more of talented people than physical capital, the Italian small-business person was among the founding fathers of 'Made in Italy', which together with 'Made in Germany' and 'Made in Japan' contributed to exemplify the international economy during the second half of the twentieth century.

Returning to the origins

'Made in Italy' is the stubborn assertion of an entrepreneurial culture with deep roots grown in the fertile soil of blue collar workers and technicians. It is the art or practice of 'making'—the heritage of virtuous industriousness in crafting and manufacturing taking place before the Industrial Revolution. Think, for example, of the city of Bologna in the seventeenth century, when its silk mills and its hydraulically powered machines for twisting yarn were at the forefront of technology before the start of the Industrial Revolution. Here was art that, when industrialization took hold, was defined by its Greek etymology as 'skill'.[1]

Since that time vocational schools and industrial factories that recruited and qualified blue collar workers and technicians acted, throughout the second half of the nineteenth century and during the first half of the twentieth century, as incubators of future entrepreneurs. After the Second World War, these individuals discovered entrepreneurial paths to fulfil their potential and multiply, thus projecting the achievements of the past into the future.

The list of the founders and pivotal path finders of 'Made in Italy'— many of them of humble origins, the children of labourers and artisans and, as were their parents, blue collar workers and artisans themselves—is

DOI: 10.1057/9781137492470.0017

very revealing. This list includes not only Italians, however. Marsala, one of the most famous Italian wines, is a product 'Made in Italy' developed by John Woodhouse, a Liverpool merchant, in the second half of the eighteenth century. This discovery involves a trail that passes through a tangle of competences and serendipity. Woodhouse was a trader who, early in the 1770s, landed at the port city of Marsala. While there he tried—and enjoyed—the local wine and decided to take some back with him to England. He added extra alcohol to help preserve the wine during the long sea journey and the resultant fortified wine proved well-suited to British taste (Ceccarelli and Grandi, 2011). As a result of Woodhouse's initiative an entrepreneurial revolution in the production and marketing of wine was started in Sicily. The British, who founded a genuine dynasty of Anglo-Sicilians, expertly narrated by British historian Raleigh Trevelyan (2002), showed that innovative entrepreneurship could flourish even in the dry and sunny lands of Sicily.

Among the Italians, the industrialist Enrico Piaggio (1905–1965) had the brilliant idea of building a low-cost motorcycle that would be very affordable for everyone. Corradino D'Ascanio (1891–1981), an engineer and inventor of the first prototype of the modern helicopter, was recruited to design the new motorcycle, the 'Vespa'. Drawing on his experience in the aeronautical industry—for instance, the design of the 'single-tube' support for the front wheel was derived from that for aircraft undercarriage—D'Ascanio delivered the right product at the right time. The case of Piaggio illustrates how much 'Made in Italy' has exploited ideas from apparently disparate sectors.

Making use of the combination of aesthetics and technology, Giovanni Borghi found a path in the business of domestic refrigerators. Offering a wide range of colours in addition to traditional white and making use of foamed polyurethane for thermal insulation were the two features in particular that led to the international success of Ignis, the company he founded in 1946, now a brand of Whirlpool. Born in Milano in 1910, Giovanni stepped into his father's laboratory of electrical equipment when he was about 13 years old: his father was a worker and son of workers.

Leonardo Del Vecchio (Milano, 1935), founder and chairman of Luxottica, global leader in the eyeglass frames industry, was born into an impoverished family and given to an orphanage by his mother. Based on the triad of 'integration, distribution, quotation'—namely, vertical integration, that has allowed the company to produce quality eyeglasses

DOI: 10.1057/9781137492470.0017

with the lowest production costs and the highest margins in the industry; a direct, strong presence in the retail business, with its brands sold in the company's own exclusive shops; and a stock-market listing in 1990 on the New York Stock Exchange to raise its accountability and image—Leonardo found a path that put Luxottica on course for rapid international expansion.

These are but three outstanding examples selected from the multitude of micro and small enterprises that have helped build the foundation of 'Made in Italy'.

'Elbow-grease' versus cutting-age vocations

'Made in Italy' is the celebration of evolutionary entrepreneurs—path finders whose life-force power is released by their vocations that 'anoint their hands and elbows with the grease for machines'. These are vocations different to those of revolutionary entrepreneurs—the path creators who first mastered science and become familiar with scientific discoveries, and then abandoned their knowledge maps, replacing them with creative ignorance. A prime example of such a path creator is Akio Morita, who was the co-founder in Tokyo in 1946 of a company that later changed its name to Sony.

A graduate in physics of Osaka Imperial University, Akio was the eldest son and heir of one of the oldest family-based producers of sake. The young Morita showed a strong inclination to pursue higher studies in applied physics as well as curiosity and an aptitude for the then nascent electronics, a passion for music inherited from his mother and, not least, a centuries-old legacy of entrepreneurship.

Italy too has had enlightened characters like Morita. Consider for example Camillo Olivetti, an engineer who attended Stanford University and in 1908 founded Olivetti, a paradigm of modern twentieth-century design in manufacturing business machinery. Camillo's entrepreneurial venture, continuing with the involvement of his son Adriano, with both their visions on distant horizons, represents an aberration on the landscape of 'Made in Italy'. The cutting-edge vision ended with Adriano's death, however. The Olivettis, as well as the inventor and entrepreneur Guglielmo Marconi, about whom there is more in *Stories of Innovation for the Millennial Generation* (Formica, 2013), opened up the paths to what was to become the digital revolution. However, the winds favourable to

DOI: 10.1057/9781137492470.0017

'Made in Italy' are the same ones which kept Italian entrepreneurship away from that revolution. The victors have been the evolutionary entrepreneurs existing closer to the visible horizon: they now reach out towards the new horizon of digital technologies that will be installed in their manufacturing operations.

The second life of manufacturing

Manufacturing survived the Great Recession and is now a leading player in the theatre of the intangible economy. But, there is more: it is experiencing a new renaissance. Productivity and big data, which extend the reach of companies in order to fulfil the needs and demands of a growing number of consumers, raise the profile of manufacturing and bring it back home to the advanced economies after years of being outsourced to emerging nations.

The extent to which productivity can increase in the manufacturing sector is illustrated by the United States and Sweden, where between 1996 and 2009 there have been increases in output of 57% and 69%, respectively. If factories in developed countries are no longer the principal job creators—indeed, many jobs there have been lost—manufacturing remains the undisputed star of productivity, R&D and exports. It is the intangible values, ideas and expertise nurtured by manufacturing which increase productivity, instigate innovation and set international trade in motion.

The McKinsey Global Institute report 'Manufacturing the future: the next era of growth and innovation' (Manyika et al., 2012) is a strong indication of the renewed central role being played by manufacturing industries. 'Manufacturing', McKinsey's analysts say,

> Generates 70% of exports in major manufacturing economies—both advanced and emerging—and up to 90% of R&D spending in industry. Driven by global competition in many subsectors, manufacturing's share of productivity growth is twice its share of employment in the EU–15 nations and three times its share of US employment. Such productivity growth provides additional benefits, including considerable consumer surplus. Manufacturing also plays a critical role in tackling societal challenges, such as reducing energy and resource consumption and limiting greenhouse gas emissions.

Italy is in the group of large developed nations that are making the ascent to global manufacturing. Italy has for decades been among the top 15

DOI: 10.1057/9781137492470.0017

manufacturing countries in terms of the share of gross value-added at current prices of global manufacturing: sixth in 1980, fourth in 1990, sixth in 2000 and fifth in 2010, at the time of the Great Recession. With regard to the value of manufacturing in terms of GDP, McKinsey places Italy seventh in the world (15% of GDP), preceded in the European Union only by Germany (19%).

What makes Italy different from other countries at the top of the rankings is the high level of labour-intensity of its manufacturing. Italy is an exception, says McKinsey, because as wealth increases the share of labour-intensive manufacturing output decreases. In the group of industries with relatively high levels of absorption of labour, Italy is among the top ten countries with regard to global market share, being in third place with 7%, preceded by China (36%) and the United States (11%). In this respect Italy is unique in the context of the advanced economies: in fact, labour-intensive production accounts for some 18% of manufacturing value-added. On the basis of GDP at purchasing power parity per capita, this figure should be substantially lower. However, there is a unique feature that must be considered—the impact of cultural heritage on the production of textiles, clothing, leather, furniture, jewellery and many other items that enrich 'Made in Italy'. Here, perhaps, is the Achilles heel of reductions in levels of employment because these are labour-intensive industries, very sensitive to labour costs and, therefore, candidates for emigration to nations where labour is cheaper. However, this can be countered by the action of path finders who foster creation and support the development of innovative start-ups in 'Made in Italy'.

A growing number of technological innovations and ancillary services arise in manufacturing with the aim of attracting some of the more than one billion new middle-class consumers in Asia who have opened up the world market. There is a long list indeed of innovations that pave the way to unprecedented business opportunities and give cause for detailed consideration of new entrepreneurial paradigms, particularly on the part of innovative start-ups. These will give new life to Italian manufacturing with, for example, components made using carbon fibres, nanotechnology, advanced robotics (Italy is in the lead group with Germany and Japan), 3-D printing, and the Internet of Things which makes objects 'intelligent' with the use of a wide range of sensors.

To give real meaning to the reasons why the interplay between new talents and innovative start-ups is so critically important, educational policies will need to make their contribution. There have been too

DOI: 10.1057/9781137492470.0017

many years exemplified by a shortage of highly educated and competent people, and by employment opportunities for low-skilled workers in steady, slow-motion decline. The garden of talents flourishes by filling the holes dug by disciplines born in the age of monolithic manufacturing, now drawing to a close. In the hybrid and fragmented age in which we are immersed, 'Made in Italy' will need to profit from its ability to distil entrepreneurial values by the convergence of various forms of knowledge: innovative start-ups can accomplish this task. As Nassin Taleb argues in his book *Antifragile: Things that Gain from Disorder* (Taleb, 2012), start-ups in a state of precarious balance on the thin wire of innovation derive benefit from the uncertainty, disorder, stress and shock caused by the Great Recession in the latter part of the first decade of the present millennium—the very opposite of the supposedly 'robust' world of monolithic manufacturing of yesteryear.

'Made in Italy' through the eyes of the Internet of Everything

In 2012, new specializations exceeded 70% of the total surplus of roughly €105 billion generated by manufacturing in Italy; in 2000 the proportion was around 40%. It remains to be seen whether the new specializations, including 3-D printing, will change the course of 'Made in Italy' with a major increase in productivity. Labour might give way to the entrepreneurial activities of former employees taking on the role of high-technology craftsmen and women. With 3-D printing technology, they would be involved in simplifying complex manufacturing processes and rapid prototyping as part of the inception phase of the innovation process. In a genuine spirit of innovation, new species of entrepreneurs would replace labour and occupy the many niches of 'Made in Italy'. A heritage of 'Made in Italy' is the Motor Valley which, from the factories of Ducati to Ferrari, from Maserati and Lamborghini to Dallara Automobili, stretches for some 80 km in Northern Italy. It is the designers and engineers of these world-famous motor manufacturers who are showing how 'Made in Italy' can be entrepreneurially invigorated by the creativity available in the use of the most advanced 3-D printing machines.

In the future of manufacturing 'Made in Italy', a leading player is the Internet of Everything (IoE), which shows that the economic and social effects resulting from smart connections have occurred with surprising

DOI: 10.1057/9781137492470.0017

speed. The connectivity which only a short time ago was no more than digitized access to information is now the digitized world—the IoE that connects people, processes, data and things. People are connected effectively; processes provide the right information at the right time to the right people; data generate information increasingly useful for decision-making; and things manifest themselves as digital devices and objects connected to the Internet and to each other, so enabling intelligence-based decisions to be made. The IoE teaches the client how to create more value and therefore provides the company with the information tools needed to produce what the customer requires. The more complex the product, the better the opportunity the company has to learn what makes the difference. With the customer who becomes a 'prosumer' (co-producer and consumer), the approach to sales and marketing changes fundamentally.

Leveraging the IoE for making Italian ice cream

Bologna relies on firms at the cutting edge of advanced mechanics, a strategic component of the 'Made in Italy'. One example is that of Carpigiani, the business founded in 1945 by brothers Bruto and Poerio Carlo Carpigiani following their success, two years earlier, in designing and building the first automatic ice cream maker. From that time until today, Carpigiani has managed to become and succeed as a market leader in the field of traditional ice cream machines for making the now world-famous 'Italian ice cream'.

Carpigiani uses the power of the IoE both to monitor all its machines around the world and carry out preventive maintenance. In fact, every machine is networked with the company and can interact with the engineers who assess the state of operation and program interventions. This meets a critical need of its large customers such as McDonald's, Ikea, Grom and others, with Carpigiani enjoying international recognition as a result.

There remains huge potential in connecting what is not. The acquisition by Google for US$3.2 billion of the home automation company Nest Labs is a sign of the growing market interest in the IoE. As outlined by Parag Gondhalekar, Director of Cisco Consulting Services, in his speech at the Innovation Value Institute Summit on 23 September 2103,

DOI: 10.1057/9781137492470.0017

99.4% of things are unconnected. 96.5% of things are consumer objects whose segment accounts for the vast majority of the things in the world. On average, there are about 200 things per person in the world. Nearly two-thirds of the total number of things are in the developed countries, although these countries account for only 14% of the world population. The value of the 'Internet of Everything' (IoE) at stake between 2013 and 2022 is around 14.4 trillion dollars of which 9.5 trillion (66%) will be generated by industry use cases and 4.9 (34%) from cross-industry use cases. The asset utilization should improve by about 2.5 trillion dollars in the same order of magnitude as employee productivity; supply chains/logistics efficiency by 2.7 trillion and customer experience by 3.7 trillion. Innovation is estimated at 3 trillion. IoE will grow aggregate global corporate profits up to 21% by 2022.

With 'things' that increasingly become objects of thought, knowledge and communication, Italian manufacturing firms, especially those smaller in size and those less connected to digital networks, both entrepreneurial and social, are faced with a sudden and discontinuous step: from 'making a hit in business' to a continued and persevering presence on the IoE cloud, so that the potential of the globalized economy can be explored and exploited. As a result of the incremental innovation that path finders have pursued, the manufacturing aspect of 'Made in Italy' has not tumbled down from the high mountain of the Great Recession. It is alive and well, not a phoenix to be reborn from its own ashes after death. It must, however, accelerate the entrepreneurial metabolism, resorting to increasingly massive doses of innovation in both digital technologies and business models.

Start-ups: swimming in the tide

Path finders of 'Made in Italy', founders of innovative start-ups, are forced to swim upstream. Against them are overvalued companies, those that privatize profits but offload their debts to taxpayers. To paraphrase Gresham's Law—'bad money follows good'—named after Sir Thomas Gresham (1519–1579), an English merchant and financier in the service of the British monarchy, it could be said that the bad companies drive out the good start-ups.

In the Ptolemaic system of Italian society, labour is at the centre of the economic universe, around which orbit other 'celestial objects', starting with companies. Thus it follows that in the dense darkness of the current economic crisis public funding is required to make labour's light shine.

DOI: 10.1057/9781137492470.0017

New business opportunities remain in the twilight zone, subservient to resolution of the labour problem.

The public hand, which opens temporary and unsafe escape hatches to the job market, is pushed by the arm of 'statism' and corporatism, engines for corruption and stranglers of meritocracy. To set start-ups in motion, there is a need for fighting spirit to declare all the relevant facts in order to give substance to a dream, to convince others, whether individuals or communities, to believe and invest in a new business idea. This is the good money which, driving out the bad, could multiply employment opportunities in 'Made in Italy' in the coming years.

In the Italian media, bad news has consigned the good news to dark cellars. Stories about start-ups are confined to those pages aimed at insiders and the initiated. In contrast, quite a different emphasis is given to rising unemployment, to the factory that closes down. That the planet 'Labour' orbits the 'Start-up' sun should be a headline on the front page. When will the media take note?

Start-ups: the challenges ahead

Robots and digital technologies are the terminators that destroy jobs. Even the Chinese need to be wary of the use of robots in manufacturing. For typists, cashiers, administrative staff in transportation and logistics, and many others, digital technology is the key that restricts and shuts off access to the labour market. In Italy, ageing of the population and penetration of digital technologies open doors, respectively, to employees to care for the elderly and to developers of applications for mobility platforms. In both cases, and especially in the second, the job is far from an immovable obstacle. The status of intrapreneur in mobility is poised to take over that of the worker. The job that changes in a non-traditional entrepreneurship raises uncertainties in many people.

Entrepreneurship needs to rise to the challenges posed by adaptability and education. An emerging élite group with knowledge maps of 'Made in Italy' has discovered entrepreneurial paths along which to travel. These are innovative start-ups inspired by the saying: 'In the new way of working, the work is not a place where you go, it's what you do, it's what you are.' This is a scenario dyed pink, but not entirely rosy. In Italy, policy makers have encouraged the creation of start-ups in the incubator environment and, with the support of the private sector, have given

DOI: 10.1057/9781137492470.0017

them initial nourishment. Because the terrain is uneven, policy makers should act as highway engineers, providing a good road surface (flexibility, mobility and autonomy of work), training experienced drivers (intra- and entrepreneurs), marking the lanes for start-ups that grow rapidly, and creating emergency exits to limit the damage of failures. The blossoming of so many start-ups would accelerate the metabolism of manufacturing 'Made in Italy', with a small number running fast acting as catalysts to accelerate employment growth.

High-growth start-ups are, as Schumpeter noted, subjects to fits and starts. They transcend the lines of industrial policy; public policies must therefore maintain the course between the rocks of Scylla (the scissors of digital technologies that cut jobs) and Charybdis (the many small businesses that offer low wages and lack of job security and are incapable of growing). Where can high-growth start-ups be found on the knowledge maps of entrepreneurship in Italy? This is the highly critical question path finders are required to answer. The genetics of Italian entrepreneurship leads to the choice of manufacturing and related services, with the ICT that gives momentum to the innovative excellence of 'Made in Italy'.

If the digital economy produces US$20 trillion in revenues, the world of atoms—manufacturing—is worth about US$130 trillion, according to estimates by Citibank and Oxford Economics. This is the world where the revolution in innovative manufacturing by artisans making high-tech and low-cost products is in progress. With 3-D printing technologies becoming less expensive, promising entrepreneurial perspectives open up for manufacturers. Italian craftsmen, labourers and technicians can design customized items on demand. Manufacturing, done at home and in factories, tailored to consumers: this is the new entrepreneurial renaissance for elevating Italian industry to 20% of GDP, as the government and social partners desire.

The re-industrialization of Italy with the rise of leading-edge manufacturing start-ups will need to involve systematic collaboration between established enterprises and a new generation of path finders. This can be achieved with higher levels of education for both employers and employees and by emerging from the knowledge silos in order to participate in networks where entrepreneurial ideas circulate freely among firms—even with competitors. It is for the path finders to promote and cultivate open innovation, to breach the dam of knowledge so that its water can irrigate the land where new entrepreneurial species can be grown. However, the water must not be spread at random: knowledge channels will be needed

DOI: 10.1057/9781137492470.0017

to ensure that each seed of ideas of an entrepreneurial species can grow and hybridize with other species. This will extend the horizon of opportunities and innovative business solutions across the boundaries of an industry sector. Those who share ideas with others completely different from themselves multiply ideas by finding connections in less obvious areas of application. Open innovation is therefore not a reserve for specialists. All employees of an enterprise are fully involved, regardless of their position and status in the organization and, with them, external actors come into play, putting forward external and independent points of view. Top management must ensure that both internal and external contributions can be absorbed by the corporate body as a whole.

MakeTank: the vanguard of the makers of 'Made in Italy'

With the creativity of the craftsman backed up by digital technologies, Italian makers are the builders of bridges linking the two sides of the past and future of 'Made in Italy'.

MakeTank is the platform of Italian makers, an initiative that will involve designers, creative people journalists and experts in the 3-D world combined with 'Made in Italy'.

As Laura De Benedetto co-founder of MakeTank says, 'What it does and will continue to make a difference in the future will be the imagination and preparation of Italian designers in designing functional objects, socially useful and also customizable and perceptible as "Made in Italy"'. Among the new products that Italian makers are bringing to market there are *Skiddi*, an innovative accessory for skiers, *Cambiami* (*Change Me*), an eyeglasses that can be rapidly changed according to their wearer's mood, and *Twist of Fate*, a range of elegant jewellery.

WIB: the vanguard of innovative solutions in the vending market of 'Made in Italy'

Retailing faces demographic, economic and technological changes, in addition to meeting the changing demands of time-starved, well-informed, connected consumers accustomed to the 24/7 convenience of online shopping. Consumers also seek better value and convenience to satisfy their immediate purchasing needs off-line, wanting nearby and accessible shops—especially outside 'normal'

DOI: 10.1057/9781137492470.0017

working hours. Concept and brand stores, in-store promotion, branding campaigns based on the release of samples and gadgets are also expanding their role: they transfer emotional value to customers and their success is increasingly related to integration with viral and social campaigns.

Traditional, physical shops appear to be unsuited to supporting new business models based on the Internet of Things paradigm because they are slow to adopt innovation and integrate off-line and online experiences. Furthermore, they require high levels of investment and incur high operating costs and are often impossible to install in unconventional albeit high-traffic areas such as corporate offices, universities, metro stations, airports and hospitals.

The survival of 'Made in Italy' in a globalized and interconnected world might require the development of tools enabling new unmanned channels that can at the same time engage the customer in a rich shopping experience. Vending systems are already part of the Italian mechatronic landscape, with many SMEs and some very large players leading the production of, for instance, coffee and snack dispensing machines. There are many competences here that could be easily moved into the new IoT arena.

These considerations supported the proposal to bring together designers, automation engineers and web designers to explore the new frontiers of automatic retailing, where innovation could provide services located near to potential customers and thus increase opportunities to generate sales.

WIB, a new hi-tech company, emerged from this experiment which was held at the ARCA incubator hosted by the University of Palermo. With help from the prototyping laboratories and the network of small manufacturing companies collaborating with the incubator, WIB has designed an innovative system for implementing stock handling, interactive user experience and management technologies; this convinced Business Angel and Venture Capital investors that it could deliver a paradigm shift in convenience and automated retailing. Intel Corporation gave the award of most innovative solution of the year in the vending market to WIB machines; while Ipercoop, one of the largest retail players in Europe, has gone into partnership with WIB to test the concept. Preliminary designs for dedicated WIB shops selling Italian brands in different sectors have been produced.

DOI: 10.1057/9781137492470.0017

In addition to onsite purchases, consumers can access WIB stores online, choosing the most convenient one from a map, buying online and then collecting purchased items at their convenience, day or night.

The WIB web-based approach and tools offer strategic advantages to retailers who can increase revenues and sales volumes using advanced sales techniques such as dynamic margin variations and combined promotions coupled with increased business efficiency (for instance, using 'out-of-stock' alerts) and generate new revenue streams and increase customer loyalty by using embedded visual devices.

Outward communication channels and an open-minded attitude to new ideas and major changes brought about by those different from ourselves are opportunities not to be missed for 'Made in Italy'. This is taking place in the wake of the sharing economy and open innovation that a fresh generation of young entrepreneurs is currently working on, in tracing the map of 'Made in Italy'. For example, the business start-up Poshman, a social network of fashion and style, seeks to achieve a complete proximity mapping of 'Made in Italy' that is closer to the consumer: this includes not only the most well-known brands and shops, but also those less well known. Its portal connects fans and fashion professionals and makes it possible to capture images of a product (clothes, shoes, and so on) and share them in a network in which each follower will know how much the item costs and where to buy it. It is, in short, a version of Tripadvisor shops, a compass for navigating the world of style and fashion. As the founders of Poshman say, 'Every user can in fact leave a comment about [their] experience of buying and on the collection presented by the store.'

The three capitals of 'Made in Italy'

Collaboration on shared values accelerates the adoption of innovations that give rise to new markets. Joint creativity ('co-creation'), creative ways of communicating and experimentation of new ventures are substantial challenges looming on the horizon, supported on three pillars: imagination capital, relational capital and entrepreneurial capital. In wanting to be innovative with others, the question to be answered is this: how can human, organizational and financial resources best be exploited?

DOI: 10.1057/9781137492470.0017

To co-create it is necessary to develop a new business image. It is not enough simply to invest in knowledge. The message of Albert Einstein that 'Imagination is more important than knowledge' must be given due attention and accepted. Entrepreneurs who do not invest in imagination capital rule themselves out of the open innovation game.

It is relational capital, with its network-shaped structure, that facilitates interactions between people, resources and ideas across organizational boundaries. Open innovation recalls the thoughts of Abbot Ferdinando Galiani, a Neapolitan economist of the eighteenth century, who argued that markets are conversations. The quality of the network of personal and business relationships—that is, the combined intelligence of people and organizations with different skills and abilities—plays an increasingly critical role for a 'meeting of minds' along the route of innovation. Business networks and technological districts that are spreading in Italy will have many more chances of success given both greater confidence and levels of investment in relational capital.

Experimental entrepreneurship for developing entrepreneurial capital is the strongest message emerging from the manufacturing revolution. In order to spread the spirit of doing business in society and in individual firms, open innovation relies on laboratories for testing innovative start-ups with high expectations of sustainable growth. The orthodoxy of the business plan developed in incubators and business divisions with the aim of seeding spin-offs drags entrepreneurial potential down. Laboratories for new venture experimentation replace the business plan with a business process development in a cultural environment in which the entrepreneurial spirit is everywhere, where anyone can experiment and try something new, where there is free access to the financial and human capital required for taking innovative ideas forward.

Imagination capital, relational capital and entrepreneurial capital are hidden in the depths of 'Made in Italy'. Once visible, it will be the result of individual and collective behaviour, not just that of policy makers, that will separate the winners from the losers.

Hopes for the future: Homines novi in the Internet of Things

The Italian economy has become bloated as a result of excessive bureaucracy and corporatism, but its brain remains agile thanks to the adrenaline

DOI: 10.1057/9781137492470.0017

of its internationally successful manufacturing industry. 'Made in Italy' knowledge maps mastered by Italian niche-based multinationals are so detailed that they can identify paths that will enable these businesses to stay ahead of their rivals in international markets. Strengthened by significant manufacturing achievements in the Peninsula, a wind of creative emulation is blowing in the direction of Germany. 'We can catch up with you and, one day, we may even overtake you', is the message blowing in this Italian wind. Shadows may loom over the contribution of manufacturing to GDP, down 5 percentage points since 2000 and now standing at 15.5%, but light emanates from the new industrial specializations in medium and high-technology, from automated manufacture to the chemical and pharmaceutical industries which, according to estimates by economist Marco Fortis, increased their joint contributions to Italian foreign trade surplus in manufacturing from 41% in 2000 to 71% in 2012.

Meanwhile, the field of creativity is expanding day by day. Driven by the convergence of scientific and social cultures, the technologies of the Internet of Things contribute to the cultivation of that creativity. In Italy, following the commercial success of the Arduino microcontroller (used to develop interactive objects), Massimo Banzi is opening up untrodden paths for manufacturing companies with the 'Officine Arduino' (Arduino Workshops) incubator. 'It came into my mind,' says Banzi, 'that Arduino could serve as an incubator of new ideas: the machines are there as is the desire to create a space where talents are gathered together to develop new products, and to rediscover their "maker gene": the impulse to learn and make things by yourself'. More new paths could also be created if Italy gave credit to the business opportunities brought to light by the serendipity of young people: sadly, this 'accidental sagacity' goes unrewarded and this neglect prevents 'Made in Italy' from achieving growth through a 'grand unification' of its two great forces of manufacturing and culture. In contrast, digital-technology experts in manufacturing and rock-music artists in Berlin serendipitously find common ground for such a 'unification'—as reported by Jeevan Vasagar in the Financial Times (Vasagar, 2014).

The courage of its industry to innovate brings Italy close to German shores, but its reluctance to promote and give voice to innovative start-ups causes it to drift back again. However, although Germany occupies a high-ranking position in the world championship of innovation, it falls short in another major competition—that of innovative entrepreneurship

DOI: 10.1057/9781137492470.0017

with high expectations and strong growth potential: Italy, though, performs worse than Germany. The Global Entrepreneurship and Development Index, developed by Zoltan Acs and Laszlo Szerb, measures entrepreneurial performance in over 70 countries and shows Italy 11 places behind Germany. Ranking higher than Germany with regard to perceiving opportunities for new ventures, Italy clearly loses ground when that perception has to be translated into implementation: achievements supported by venture capital seem few and far between, evidence of which is the relatively small amount—€100 million—invested in innovative start-ups in 2013.

The success of Italian manufacturing abroad indicates that export-oriented SMEs and companies seeking international markets improve production processes, reduce the costs of components, enrich existing products with new versions and introduce new models. There are strong and relentless pressures for innovation that will meet customers' demands today, but not for innovations that can break through the wall behind which the latent needs of the customers of the future are concealed. In short, Italy suffers from a deficit of non-incremental innovations—those innovations that are associated with start-ups and the generation of new products and services that did not previously exist. Companies already well established in the markets face a dilemma: if they put their foot down on the accelerator of efficiency, the rush to innovation slows. The drive to do better what the company already does well constrains innovation within the boundaries of the visible horizon rather than leaving it open to what lies beyond. Effecting disruptive innovation in an existing business is a difficult and controversial task.

There is no better card than creative ignorance that the imagination of the stereotypically 'flexible' Italians could play against the apparently perfect knowledge of the stereotypically 'rigorous' Germans. It is, then, imperative that Italian manufacturers provide support to their *homines novi*, entrepreneurs of game-changing innovation in the Internet of Things. These should include scientists in fields ranging from information technology to biotechnology. Using their ingenuity, scientists could enrich the technological image of 'Made in Italy' with new content. For this to happen, the policy of the Italian universities should be to foster entrepreneurship among academics in scientific specialities. The strong cultural imprint of 'Made in Italy' would secure new lifeblood in the interaction between teaching, research and innovative start-ups by academic scientists.

DOI: 10.1057/9781137492470.0017

The coming of a new age: finding paths of digital tourism

With an environment rich in natural and artistic beauties perhaps unique in the world, Italy has been greatly enriched by acquisitions from other civilizations that existed for many centuries. Finding paths of digital tourism could now bring their knowledge maps alive.

Together with manufacturing, tourism is potentially a major exit route from economic stagnation in Italy, with digital interactivity for cultural attractions the vehicle and entrepreneurial champions of tourism with cultural DNA the drivers behind the wheel. Investments in the digital arena and culture will be the fuel, the former because 'cybernetic tourism' already exists as a result of the use of ICT in a variety of ways, including the Internet, mobile phones, tablets, and so on. By using digital technologies, numbers of cyber tourists are increasing, with those from China at the forefront. Cultural investments will fuel the vehicle because Italian tourism is—or rather should be—based on the nation's culture. If the marriage between technology and culture threatened to fall apart, that route out of the stagnation would be blocked. After the wedding, the Italian model of tourism would be redefined, not simply reformed. In short, Italy would outstrip competitors not by improving its tourist routes, but by designing entirely new ones. Pathways of exploration, storytelling, games, and interactive participation would boost both the volume and the quality of tourist numbers in Italy. A strong evolutionary component of leisure and cultural activities would take visitors to the 'fair country where the *sì* is heard' (Italy as described by Dante in *Inferno canto* XXXIII, verse 80).

Technology can deliver information directly into the field of vision of those who make full use of portable digital media. For example, wearing the Google Glass enables tourists to enter enhanced reality; with Oculus Rift they are immersed in virtual reality. The culture gives them the long view of Lynceus—the Argonaut (perhaps a precursor to the modern-day adventurer?) who is said to have had excellent vision.

Such a scenario would create a momentum of interest of international tourism in the Italian cultural heritage. If this is the perspective in pink, today it is still tinged with grey. The cultural offering is often limited to traditional guided tours of cities and museums, and individual cultural goods, or otherwise to the mere exploration of sites that are not well guarded. The attraction for the tourist comes from a

DOI: 10.1057/9781137492470.0017

mythological perception of the cultural attraction in its own right. During the months in which the remains of Pompeii were crumbling before the eyes of visitors, the Louvre in Paris replaced traditional audio guides with multimedia portable gaming consoles (Nintendo 3-DS) with small, three-dimensional screens. This choice had nothing to do with a lack of visitors; rather, it was because the need to improve the service, using the latest digital technologies, had become obvious by virtue of the substantial and proven advantages of digital media with regard to learning processes.

Italy is an open-air museum. Let us put the closed spaces of the famous Parisian museum behind us and let us imagine innovative solutions for open spaces. Consider the case of the Appian Way (*Appia Via*) of which Marcus Tullius Cicero (106 BC–43 BC), the great master of the Latin language wrote in his *Oratio Pro Tito Annio Milone*: 'nunc eiusdem Appiae nomen quantas tragoedias excitat!'—'Now what tragedies does the name of that same Appian Way awaken!' (Smith Purton, 1886). Thus from this tragedy (there had been bloodshed on the Appian Way) to the charm of ruins, more than a thousand years old, of the most famous road in the Roman world, hybridization of digital technology and culture helps visitors interpret the ancient items, events and ideas from the distant past brought to their attention.

For an entire school class, that experience might sound like an adventurous treasure hunt, led by 'Travellers into Time'; while the most enthusiastic experts in archaeology might wish to buy vintage miniatures, made before their eyes by a 3-D printer. The experience could be further enriched with holographic performances—in full size and without the need for glasses—in the most meaningful and evocative locations. Visitors would be transported into ancient times, immersed with their senses and through narratives in the customs and habits of the historical period. Celebrities of those times could react and respond interactively to questions and comments, with concealed actors bringing the holographic characters to life with the use of Motion Capture technology. The Appian Way would become an ecosystem of information, leisure and cultural contents. Tourists could take advantage of a full range of services—for example, cultural tours, events, information about mobility, instant voice translation and mapping of hotels and restaurants.

Tourism that marries culture and technology makes the visitor an interactive participant of a renewed Grand Tour, following in the wake of Johann Wolfgang Goethe, who set out on a journey to Italy between

DOI: 10.1057/9781137492470.0017

1786 and 1788 and then described his experiences in *Italianische Reise* (Goethe, 2010), published in 1816–1817. The combination of culture with technology makes such a journey constructive and fun by harnessing the emotional power of the film industry, the sense of immersion given by virtual reality, the widening and deepening of the vision with the enhanced reality and the high level of interactivity of video games.

The journey therefore places visitors of historical sites like the Appian Way in time as well as space, because cultural tourism always has time embedded in it. Landscapes, buildings and characters are faithfully reconstructed. In setting the path out of economic stagnation, more important efforts lie ahead as far as the entrepreneurial champions— who should have no difficulty in following that path—are concerned. So, this is where the connections between research and innovative entrepreneurship come into play. Humanities will first need to prove their ability to think and act as a transmission channel of knowledge directed to entrepreneurial ventures which, in turn, facilitate and secure the journey all along the route where tourism meets and combines technology and culture.

Tourism in Sardinia: a highway full of crosses and without digital delights

Is it possible for the land of *L'Unione Sarda*, the first European newspaper to adopt an online edition, and Tiscali, one of the first followers of the digital age, to start a revolution in tourism?

The coasts of Sardinia are wonderful for a summer holiday 'To.o' ('Tourism zero dot zero', not Internet). Sun, sand, sea; reading for recreation, not work—maybe a good book. Do you intend to take a leap into 'T1.o', having a love affair with digital experiences outside the work? If you succeed, do not harbour any illusions, though. The network connection is a flimsy and powerless inspiration, as if it was a dying breath. Do you chat or dance together on Facebook? No problem. From the raffle of connection sooner or later you will find a winning ticket. You are connected! And shortly thereafter, disconnected; then reconnected; a cycle that does not harm the dance. What is more, how do we hold a conversation face-to-face at the seaside? Two words; a pause for diving into the water—it is so hot; two more words; nothing different from the other cycle.

DOI: 10.1057/9781137492470.0017

It may be the case that among vacationers there is some bizarre character who wants to make a double somersault at T2.0. This person indulges in cultural idleness. Removed from the tyranny of the desk and oblivious to the daily routine at work, the vacationer plans to explore the world of culture holding a paper book with a superb landscape in front of her eyes (let us make her female). She's reading about Epaminondas and Pelopidas, the Theban heroic leaders. She is fascinated by the battle of Leuctra (371 BC); having left the printed encyclopaedia at home, she attempts to use Google to find out more—but discovers there is no or close to zero connection to the Internet. After the holiday, back at the desk, the pressure of other matters relegates Leuctra to the attic. 'Cultural idleness' is an oxymoron along the golden coastline of Sardinia. Leisure there and culture at home; the recipe for hybridization is unknown.

The pre-digital age and pre-virtual reality are together a shroud of the 'good old days' that obscures many tourist landscapes in the Italian peninsula. Fixed in the pre-digital age, Sardinia, as with other regions with extremely high rates of youth unemployment and with young people neither in work nor in education, employment or training, places work and not entrepreneurship on the throne of the expectations and ambitions of its young citizens. This is a mortal sin because innovative entrepreneurship is, particularly in tourism, the only source of employment. Culture is the DNA of Italian tourism. To release it and make it visible to the naked eye requires entrepreneurs able to demolish the membranes of a tradition of hospitality that has become obsolete. There will be a great leap forward for tourism in Sardinia once the hidden entrepreneurial spirit of young people of the digital age is discovered. One-dimensional tourism can give way to a multi-dimensional version where physical and virtual realities combine and where added value can be achieved through their interactions.

Imagine Sardinia gifted with the long view of Lynceus and wearing Google Glasses to bring virtual reality into its own visual field. Imagine the attraction of its cultural sites enriched by digital media that would plunge tourists into the realities of the time lived in those places: T0.0; T1.0; T2.0. However, holiday tourism does not end here. Lynceus with Google Glasses would see that the nomads of knowledge, migratory birds of start-ups driven by science and technology, fail to respond to his call. They love beautiful places, meet the most diverse cultures and people, mingle with them to the flowering of global start-ups—firms without either geographical borders or cultural, religious, racial or gender

DOI: 10.1057/9781137492470.0017

barriers. Many among them are the young generation of the Millennium, the digerati 3.0. Imagine a flock of this species of migratory birds landing on the rooftops and trees of Santa Teresa di Gallura, the Bosa, and other resorts of the island. They carry a litter of innovative start-ups that arouse curiosity among their Sardinian peers. There is for them a renaissance of entrepreneurial and job opportunities. Local culture marches in step with the global culture of the knowledge nomads. Currently such a scenario unfortunately remains a complete fantasy, because the cultural attitude prevails in the community that leads one to say, 'After all, the Internet is a thing, it is not life, you can do without it.' Viewing the Internet galaxy from observatories, in Sardinia and elsewhere in Italy, the knowledge nomad exists as a star hitherto unknown to the vast majority of the expert astronomers.

Keep beauty alive

As Aristotle stated, 'A great city should not be confounded with a populous one.' No other places in the world are so great by reason of their beauty as the many small- and medium-sized towns and the thousand and more villages of Italy. Travelling between one such place and another it is easy to behold in wonder so much beauty. 'Beautification' is the name of the current of beauty that throws light on the aesthetic values of those towns and villages, enhancing their knowledge. 'Knowledgefication' is the name we shall give to the current of knowledge that arouses love for beauty. The two currents together promote a harmonious social and aesthetical order that enhances the quality of living. Policy makers and civil society have a responsibility to keep intact the beauty of the Italian cultural heritage and its landscapes, relentlessly pursuing the objective of providing power to these two currents.

Wonderful places to visit certainly attract tourists, but what matters most is the attraction of the right kind of citizens and the best talents from all over the world. Traditional and digital tourism on the one hand and, on the other, brain circulation (mobility in a physical sense that stimulates face-to-face communication) and brain waves (mobility in a virtual sense that takes advantage of new, open-space technologies) are reflective destinations capable of arousing emotion in a game of crossed eyes. It would be a manifest error of appreciation to consider digital tourism and brain waves as an alternative to the decay of beauty

DOI: 10.1057/9781137492470.0017

caused by feeble-mindedness, indecisiveness and wrong and inefficient decisions.

When the weeds of decay have rooted and start to grow, those who want to uproot them should think big. When the thinking is not big, what happens? In Bologna, one of the most beautiful cities in Italy, the City Council thinks so small that the weeds of decay have taken root in the porticoes and on the walls of the historic buildings and all the houses of the historic centre, one of the largest in Europe; but the porticoes, some 38 kilometres in all and built throughout a long historic period, are an extraordinary cultural asset that is a candidate for nomination and approval as a World Heritage Site by UNESCO.

Not wishing to marry Mr Common Sense, the City Council of Bologna cleans the various graffiti on the porticoes and walls during the daytime, so that during the night the 'coarse soldiery'—similar to that described by Alessandro Manzoni in his historical romance *The Betrothed*—may have clean surfaces on which to scrawl. If he could marry Ms City Council, Mr Common Sense would find the path to undertake the work of prevention (civic education) and repression (night patrols). Entering into the marriage is, alas, something they must not do. Well then, let the owners of apartment buildings pay for the 'doing' of the coarse soldiery that scribbles on the walls and the 'undoing' of those who paint over or remove these linguistic eyesores. We may think that an annual subscription for 'cleanliness guaranteed', perhaps in the order of €130 to €200 per building is a small amount; indeed, in monetary terms it is very little, although it would still be a hidden tax that would tend to hide the lack of care and the inefficiency of the public authority. It is more if we open windows on the subject of property rights.

These are the property rights that must be the sacrificial lamb on the altar of non-common sense. Everyone can benefit from a graffiti-free view of the porticoes, palaces, monuments and many other public and private artefacts that decorate the city spaces. Beauty is an indivisible good: its consumption by one person does not reduce the amount available for others. It is this indivisibility that the coarse soldiery has thrown away to rot. Beauty is also a non-exclusive good—it is difficult, if not impossible, to exclude anyone from enjoying the aesthetics of the city. The succession of beautiful buildings in the historic centre and the uniqueness of the long stretches of the arcades are both a gift that the people of Bologna have received from their ancestors and a very pleasant surprise for tourists. The coarse soldiery acts as a maverick, free to raise

DOI: 10.1057/9781137492470.0017

a wall of ugliness that detracts from and even removes the beauty from our vision. In short, that soldiery is a group of individuals highly motivated to take action that results in divisibility and exclusion in the city.

Under these conditions, subscriptions and fines are the outcome of an asymmetry between the apartment building owners who pay and the victorious soldiery, given that the public authorities were obliged to raise the white flag of surrender on the two sides of prevention and repression. The maintenance of beauty is a very long journey. Who undertakes it should be determined and possess the qualities of a marathon runner— including the necessary psychological motivation. This is not the case with the municipal government of Bologna. The City Council is an athlete for sprint races only. During the day the municipality attempts to clean the walls; during the night there is a return to the implacably ugly. There is a pause; then, again, another short race. The local government can attach a price tag for the fight against graffiti, but this cannot be done with beauty, an intangible good whose intrinsic characteristics of uniqueness keep it outside the bounds of economic calculation. Beauty, however, gives value to the reputational capital of the city—the set of values and social behaviours that distinguish those cities that make progress from those in decline. An anti-graffiti subscription will certainly not increase the reputational capital of Bologna.

A bitter conclusion: because there is no clear sign for a path that leads to beauty, I wonder what Stendhal, restored to life, would say today. Visiting the city, this nineteenth century French writer jotted down these words on 28 December 1816, then included them in his book *Voyages en Italie* which was first published in 1826:

> 'En général, les portiques de Bologne sont loin d'être aussi élégants que ceux de la rue Castiglione, mais ils sont Lien plus commodes, et mettent parfaitement à l'abri des plus grande pluies' ['On the whole the porticoes of Bologna are far from as elegant as those of the rue Castiglione, but they are much more convenient and provide perfect shelter from the heaviest of rain.']. Stendhal, *Voyages en Italie*, 1826

The story of the beauty of Bologna being in a state of decay shows that the interplay between beautification and knowledgefication is a precondition for finding paths that infuse new vitality into the body of great places. This cannot happen if a community, its policy makers and individual citizens stay locked—seemingly in relative security—in the silo mentality that obscures the vision of possible futures.

DOI: 10.1057/9781137492470.0017

Note

1   The ancient Greek word 'tekhni', commonly translated as 'art', more accurately means 'skill' or 'craftsmanship'.

DOI: 10.1057/9781137492470.0017

10
Portrait of Path Creators

Formica, Piero. *The Role of Creative Ignorance: Portraits of Path Finders and Path Creators.* New York: Palgrave Macmillan, 2015. DOI: 10.1057/9781137492470.0018.

▶

DOI: 10.1057/9781137492470.0018

The creative ignorance of path creators accompanies their broad-minded perspective, not forced into the rigid corset of acquired knowledge. They exhibit courage in their will to look at multiple views on things beyond the visible horizon. Disobeying the fundamental tenets dictated by knowledge, path creators cast their gaze well beyond the visible horizon of certainty so that the skyline of doubt can be grasped.

Reserved knowledge constrains the path finders within clearly defined limits of relationships. Path creators start the journey alone, but along the way the unrestricted character of creative ignorance gives them freedom to forge unlimited relationships of familiarity and intimacy with people very different from themselves.

The days in which skilled artisans far from the inner circles of the most knowledgeable persons intruded on the field of path creation are consigned to the past. The lesson of the craftsman John Harrison who solved the difficult problem of measuring longitude accurately, was significant. From now on, the most varied protagonists, fugitives from the today's mainstream systems of education and entrepreneurship, will crowd that field.

'*Ignorance is bold and knowledge reserved.*' (Thucydides, Greek historian and Athenian general, c. 460–395 BC)

The world beyond the visible horizon

In stark contrast to the path finders, path creators do not simply accept knowledge as a truth. Adept at using a trans-disciplinary approach to venturing, they rely on a series of iterative experiments to observe the world beyond the visible horizon. In the words of Taleb (2012), path creators, contrary to path finders, are 'antifragilista' who 'do things without understanding them or with an incomplete understanding—and do them well'. They derive much more benefit than harm from shocks and random events. Discontinuity, randomness, breakthrough and disruption are the elements that path creators place side by side with the column 'Innovation'.

Who creates a path may be represented in the guise of René Magritte's 'The Ignorant Fairy', a portrait which shows a person who represents the promise of new knowledge. The creator intentionally goes over the path already travelled so many times, stops seeing his visions and perspectives

DOI: 10.1057/9781137492470.0018

through the judgments of others and fails to consider significant questions of others regarding the 'why' he opens doors and pursues alternatives to well-trodden paths.

Path creators—a minority more or less aggressive depending on the institutional, entrepreneurial and social contexts within which they act—pay no attention to, or if they do they disobey, the trends plotted with the aid of knowledge maps. For them, innovation is, in fact, disobedience. Therefore, noncompliant innovators do not fall into knowledge zones densely populated with past data. The lenses they wear are not intended for predicting a future that is a projection of the past: their lenses, once worn, enhance the imagination. The future is no longer a particular place identified by exercises in logic and anchored to the past. With imagination, the future is everywhere—a future revealing uncertainties not quantifiable in terms of manageable risks on the basis of the knowledge maps mastered.

Portrait of a path creator: *The Geographer*, by Johannes Vermeer (1632–1675)

Navigation charts and maps that reflect the state of knowledge of the world in their day surround geographers. Before Vermeer painted him, this geographer was probably immersed in calculations for plotting courses on those charts. But at the time of his portraiture the geographer has, intuitively it seems, turned his gaze elsewhere, to the window that opens onto an unknown landscape.

Perhaps, just outside the window, an idea is in flight that would like to enter the room, to be seized by the geographer. Here is how he began a thrill-seeking journey, a special ride to a world that knowledge maps still ignored.

Vermeer himself was a path creator, making use of a *camera obscura* ('darkened chamber'), an optical device that helped in representing reality in an objective, impartial manner.

DOI: 10.1057/9781137492470.0018

The Epicurean 'disobedient': descendant of Epicurus and Lucretius, and followers of Jean-Baptiste Say

Epicurus and Lucretius introduced the notion of '*clinamen*' to indicate the possibility of atoms swerving away from their trajectory. Path creators are their descendants. With a wide range of vision, they proceed on barefoot—that is, little or no monetary and relationship capital and rejecting current understandings and past experiences—along a course that swerves from the route traced by today's knowledge. This '*clinamen*' exhibits the characteristic of being unpredictable and acts to undermine those who work within the horizon of certainty. The paths they create trace new trends that dramatically change the direction of an industry.

For example, consider path creators like Apple and Google that have redirected technology spending towards end consumers, leaving path finders like BlackBerry struggling to keep pace. Trying to catch up, they are in a zone path finders find very uncomfortable. As noted by Richard Waters in the Financial Times:

> Companies missing out on the stock market party include some of the biggest suppliers of corporate technology, including IBM, Cisco, Hewlett-Packard and

DOI: 10.1057/9781137492470.0018

Oracle. Microsoft and Intel have also been left behind by the rise of mobile. Surging growth in emerging countries bailed out their ageing product lines for a while, but even that support has now gone. Taken as a whole, the revenues of these six tech giants actually went into reverse in their latest financial years, falling 1.5 per cent. A lacklustre recovery is projected to turn that back to growth of only 2.5 per cent by 2015. This represents a significant change in their prospects. In the four years to 2010, their combined revenues grew 30 per cent despite a blow to demand from the 2008 financial crisis. The following four years, by contrast, are expected to show growth of only 9 per cent. Making the most of this transition requires inventing new applications and mastering new business models that were not possible before. Simply trying to play catch-up against Amazon, with its significant scale advantages and tolerance for low profit margins, is not a comfortable place to start. (Financial Times, 2014)

From a different angle, as would be the case for someone who was, say, on the side of the moon facing the earth, path creators observe that the knowledge map is in motion and orbits the 'sun' of insights unconstrained by the knowledge of why something can't be done. This change of the point of view is the factor that pushes path creators to create new paths outside the map rather than find paths within its borders. From the economic science perspective, path creators are followers of Jean-Baptiste Say, for they feed the supply side of a dynamic economy by proposing innovations that involve revolutionary changes in industry, in business and in the markets, giving consumers substantial benefits compared to the products and services currently available. The supply of such innovations creates its own demand because consumers are generally inclined to spend rather than save in the presence of novel and exciting products and services.

Following the thoughts of Karl Popper, according to whom 'future knowledge is not possible in the present' (Popper, 1957), it is path creators who move towards the horizon of doubt, fortified by the strength of uncertainty. As far-seeing pioneers, possessed of resilient personalities, who create innovative entrepreneurship through discovery and rebellion against conformity in business and even in science, they appear well-equipped to take a leap into the darkness of the unknown.

The '*clinamen*' of John Harrison

The inquiring mind, if synchronized with knowledge as is the earth with the moon—that is, in an almost perfect way—does not perceive creative ignorance. The ability to act irrespective of formalized

DOI: 10.1057/9781137492470.0018

knowledge, as in the case of the carpenter John Harrison, and the ability to invent mental models outside of the fundamental tenets of physics, as in the case of Einstein, disable the perfect synchronization between the inquisitive mind and the knowledge accumulated through education, research and innovation.

The tercentenary of the Longitude Prize in 2014 invites us to consider the person of John Harrison (1693–1776), the English carpenter and connoisseur of clocks who, by building a series of highly accurate and reliable maritime clocks, succeeded in providing the means for the accurate determination of longitude. As Dava Sobel's bestseller *Longitude* (Sobel, 1995) shows, the inquiring mind of the artisan Harrison repeatedly ran into brick walls of advice and solutions raised by the most knowledgeable astronomers of the day. As Sobel writes,

> Renowned astronomers approached the longitude challenge by appealing to the clockwork universe: Galileo Galilei, Jean Dominique Cassini, Christiaan Huygens, Sir Isaac Newton, and Edmond Halley, of the comet fame, all entreated the moon and the stars for help. Palatial observatories were founded at Paris, London and Berlin for the express purpose of determining longitude by the heavens.

The English Astronomer Royal Nevil Maskelyne (1732–1811), one of the commissioners in charge of awarding the Longitude Prize (established in 1714 by an Act of Parliament), was vehement in his support for these stars of astronomy. Thus it was a long, lonely journey, littered with obstacles artfully placed by scientists, that Harrison undertook to create a path that would, centuries later, have taken mankind to the moon. According to Sobel, it seems that the astronaut Neil Armstrong, when dining at 10 Downing Street, official residence of the British Prime Minister, 'raised his glass to Harrison as "the one who started us on our journey"'.

It took 59 years before it was officially concluded, in 1773, that John Harrison, the carpenter and passionate of clocks, had truly solved the problem of measuring longitude. During this long period many ships, with their cargoes, were sunk and many lives were to be lost at sea. Here indeed was a demonstration of the high price that may be paid when knowledge presumes to obscure the great achievements of creative ignorance.

DOI: 10.1057/9781137492470.0018

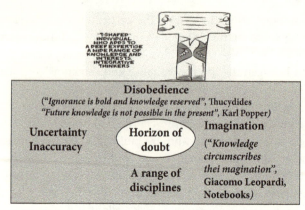

Disobedience
("*Ignorance is bold and knowledge reserved*", Thucydides
"*Future knowledge is not possible in the present*", Karl Popper)

Uncertainty
Inaccuracy

Horizon of
doubt

Imagination

("*Knowledge
circumscribes
thei magination*",
Giacomo Leopardi,
Notebooks)

A range of
disciplines

FIGURE 4.1   *The Epicurean 'disobedient'.*

DOI: 10.1057/9781137492470.0018

Path generation by practising purposeful ignorance

Happily ignorant of tradition, path creators break down the knowledge patterns they have followed in the past and, in doing so, shift their behavioural trajectories from a knowledge-burdened to a clean-sheet approach to exploration. Like explorers setting out on voyages into uncharted waters, path creators trace new paths by cultivating imaginative visions and walking randomly. Their conviction to succeed in building new paths from a void is strong and without fear, because path creators are not aware of how, and how many times, their imaginative ideas might simply disappear without trace. Freedom of action deflects them from the 'optimum' of knowledge laws and rules. 'There is no map, and charting a path ahead will not be easy', as Mr Jeff Bezos, Amazon's founder, said after he bought the Washington Post.[1] Path creators learn from the narratives of their past experiences that there is no need to learn from them. Rather than predicting the future, they help shape it by resorting to 'madcap schemes' (Westlake, 2013).

Ignoring is a mindset that has deep roots in the centuries-old history of humanity.

It is the 6th of July, 371 BC, and we are on the 'Orchestra of Ares' plain, observing the Battle of Leuctra. Epaminondas and Pelopidas, the Theban army commanders, refuse to adopt the traditional deployment of troops and, with ingenuity and courage, prepare an unconventional battle plan. Tradition dictated that the most skilled warriors were employed on the right flank of the line of warriors, because the left flank was thought to bring bad luck. In this way the best troops of the two opposing armies never clashed directly. Epaminondas and Pelopidas overturned tradition by deploying their most skilled warriors on the left flank. Their plan created surprise and panic among the Spartans and their allies, who suffered defeat (Scott, 2009).

Moving from military to business history, some 2,000 years later, it is the second of August 1968 AD: Dr Robert Noyce and Dr Gordon E. Moore decide to ignore the burden of administrative work preoccupying them at Fairchild Semiconductor, a company of which they were among the founders. Starting from 'the idea of getting back in the laboratory...to do some extended technological work before coming out with product areas that none of the manufacturers are supplying', Moore and Noyce design an alternative and innovative scheme that was to take them

DOI: 10.1057/9781137492470.0018

BATTLE OF LEUCTRA, JULY 6, 371 BC
SPARTANS

SPARTAN MERCENARIES

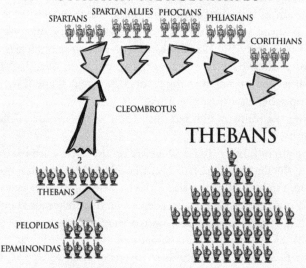

APPLE VERSUS BLACKBERRY, 2000S AD

BLACKBERRY CONSIDERED A SIMPLE TOY THE
IPHONE THAT APPLE INTRODUCED IN 2007

DOI: 10.1057/9781137492470.0018

away from Fairchild (Scandling, 1968). The future would smile on Intel Corp, their new company, rather than on Fairchild.

However, the future becomes grey when yesterday's path creators turn into today's finders.

Not-knowing is the engine that drives the exploration process aimed at creating a path whose probability of success is not something objective but, rather, a feeling, a guess or a belief inside the path creator.[2] Those who generate paths look at the true horizon which, from the path finders' locations, is obscured by 'trees' of consolidated knowledge and 'mountains' of long-standing prejudices. Path creators use the mental tool of purposeful ignorance, comparable to the echo sounder used for exploration of deep oceans, to grasp the innermost needs of the people. By purposeful ignorance, we mean that state of open mindedness which challenges what according to current views appears to be irrefutable truths. In this perspective, purposeful ignorance is 'knowledgeable, perceptive, insightful' (Firestein, 2012)—a learning process for cultivating a fertile land seeded for growing abstruse questions which reveal untouched paths. So, path creation is a singular spectacle of daring which is the fruit of 'learned ignorance'. The creation starts in the void of unknown questions. To trace that type of path empty minds are needed—those who have emptied the basin of survivor's/winner's bias, past experiences and external experts' points of view (Tjan, 2010).

Path creators and trade routes: flourishing together

Path creators generate fresh business ideas that stand out from the crowd and travel easily along new trade routes which, in turn, thrive on those ideas that are converted into smart companies. In China, during the Han dynasty, the Silk Road and the invention and production of paper flourished together. Later, under the Mongols in the thirteenth and fourteenth centuries, Samarcanda with its central position on the Silk Road was the cradle of an early papermaking establishment whose output, spread by traders along the route, supplanted other writing materials. Stimulated by the invention of paper, it was on the Silk Road that path creators would develop printing technology.

In the fifteenth century, in Europe, the combined influences of the river Rhine, a primary trade connection through Europe, and Johannes Gutenberg (c. 1395–1468), a technical genius with a visionary

DOI: 10.1057/9781137492470.0018

entrepreneurial mindset, gave rise to the Printing Revolution: printing using moveable type.

Trade routes facilitate the combining of ideas otherwise separate and distant in concept and physical location. The fact that the design of Gutenberg's printing press emerged from the technology of the screw-type wine presses of the Rhine Valley is evidence of this.

If, then, trade routes are extremely efficient ways of transmitting innovative values, restrictive practices imposed by politicians, religions and guilds slow down and impede path creators or, even worse, close their access to those routes; and how many times has this happened in history? In medieval Europe, the strict rules of the Florentine guild of wool merchants imposed a ban on exercising the profession outside of the Florentine state and enforced strict compliance with the rules established for each stage of product processing, including the tools used.

From the monasteries, Gutenberg brought out texts copied by the amanuenses. The printing press replaced their handwriting tasks almost entirely, reproducing mechanically multiple copies of printed texts. However, there are experts who still live in a cloistered environment, each of them too confined in their knowledge 'claustrum' to seek the quest of a path.

Path creators of social streets

Path creators are entering, with a great deal of determination, the social field—a vast plot of land owned almost exclusively for centuries and in many countries by public authorities.

If necessity is the mother of innovation, the midwife is creative ignorance neither having nor knowing precedents on which to grasp. In times of stag-deflation (low growth with deflationary pressures), necessity bears in her womb and creative ignorance helps give birth to the sharing of products and services for which the trend is growing rapidly: according to the Knight Foundation, an average of more than 36% per annum between 2009 and 2012. In Italy, 13% of the population is already involved in collective consumption brought about by the collaborative economy. Its occurrence leads ordinary people to fertilize necessity in a creative way. This is how, in Italy, a path that leads to the 'social streets' movements has suddenly arisen seemingly from nowhere.

DOI: 10.1057/9781137492470.0018

In Bologna, the residents of Via Fondazza, a street in the historic centre of the city, gave birth in September 2013 to the first social street in the country, and one of the first in the world. On 30 December 2013, *Struggles in Italy*, an online media, reported the news in an article headed ' "Social streets" and the mutual aid economy', stating:

> At first, the social street project's aim was modest. Federico Bastiani [the path creator] had grown up in a small town in Tuscany where people knew one another and supported each other. When he moved to Bologna he didn't like the mix of mistrust and indifference which can characterise neighbourhood relations in big cities. To try and create a sense of community, Federico opened a Facebook page called 'Residents of Via Fondazza—Bologna' and promoted the initiative through leaflets.... The residents of Via Fondazza soon started to explore the network's potential to improve their everyday life. There have been dozens of activities around the sharing of information, time, knowledge and goods. Initiatives so far include giving free piano lessons, lending washing machines, providing tips to newcomers about services in the city, giving away leftover food when going on holiday, holding street birthday parties open to all residents, and discounts for residents at the local cinema.

Today, following the example of big cities, there are many social streets. Using the keys of mutual trust and commitment not to burden the streets with too many families, ongoing experiments open the doors to access and pooling of goods and services rather than to their property. If the sharp blade of the current economic crisis cuts back salaries and pensions, both families and individuals will be able to tap into the source of collective consumption, riding the wave of sharing. Thus the inhabitants of social streets are given opportunities to use otherwise spare capacity and idle resources. A triple helix of neighbourhood collaboration, the digital revolution and social networks turns on the engine of this collective consumption, helping at least in part to compensate for the change in individual circumstances.

Where the invisible hand of Adam Smith—the 'providence' of the free market—does not reach, could the sharing economy succeed? Will it be no more than a passing fad rather than a social earthquake that redefines lifestyles and consumption? The question remains open. What is certain is that in the social street experiments the effect of mutual aid in igniting the engines of civic innovation is clearly visible. It is with the qualities of responsibility, accessibility and responsiveness that the path creators of social streets offer challenges to city governments. As the Mayor of San Francisco's Office of Civic Innovation has argued, great ideas can come

DOI: 10.1057/9781137492470.0018

from anywhere. The task of the local authorities is to be accessible as a platform for the social streets that improve the lives of their citizens.

Civic innovation unlocks new areas of liberty to transform solutions developed by the citizens into products, services, new firms and jobs. The very fact that social streets like Via Fondazza have international echoes shows that there is sufficient social energy to contribute to the formation of a broad ecosystem of civic innovation with a strong entrepreneurial impact. Furthermore, it must be remembered that trust in others and a willingness to be accepted by and associated with the others find allies in the technology that enables us to enjoy more readily an increasing share of resources. The interplay between technology and sharing drives us to have fewer things, but to use them more and better by putting them in common use with our neighbours. The collaborative economy, as well as supporting neighbourhood business activities, helps us to weave new social networks that tighten and strengthen the mesh of social streets. Going beyond the visible horizon, the path creators of social streets have given rise to a movement that will have a major impact on neighbourhood relationships.

Notes

1   As reported by the Financial Times, 'Washington Post sold to Bezos for $250m', By Barney Jopson and Andrew Edgecliffe-Johnson, 5 August 2013.
2   In accordance with the line of thought of Bruno de Finetti (1990), the founder of 'personalist Bayesianism'—see Fuchs (2010).

DOI: 10.1057/9781137492470.0018

11
Looking for Path Creators of the Grand Unification between Manufacturing and Culture: The Case of the Opera Houses

Formica, Piero. *The Role of Creative Ignorance: Portraits of Path Finders and Path Creators.* New York: Palgrave Macmillan, 2015. DOI: 10.1057/9781137492470.0019.

> Do manufacturing and culture live in two separate and irreconcilable
> worlds—manufacturing in the world of things and culture in the world
> of ideas? Is manufacturing called upon to solve production problems,
> with culture pronouncing on 'chief systems' as in Galileo's 'Dialogue
> Concerning the Two Chief World Systems'? This is a common vision of
> those who identify manufacturing with making and culture with think-
> ing: the manual labour of artisans and technicians as opposed to the
> intellectual work of professors and scientists. As a result, this fault line
> fails to recognize that the factory is a culture that can and should go hand
> in hand with academia and research centres: you will otherwise have two
> half-cultures that do not make a single, whole culture.
>
> One need only look in detail at the staging of an opera—a seemingly
> extreme example of a cultural embrace—to see how much that vision
> is little more than a stereotype. With digital technologies now employed
> as a spider spinning its web of the Grand Unification of manufacturing
> and opera, the concept of these two worlds embracing each other is now
> acknowledged and accepted.
>
> 'Two half-truths do not make a truth, and two half-cultures do not
> make a culture.' (Arthur Koestler)

Intertwining relations between manufacturing and opera houses

By setting aside their respective knowledge maps, manufacturers
and lyricists prepare themselves for this Grand Unification between
manufacturing and culture, revealing interesting approaches for future
growth influenced by path creators. Opera houses are a component of
culture whose intangible assets—imagination, relational, reputational
and entrepreneurial capital—are, like atomic particles, invisible to the
naked eye. These are assets that remain when, after the performance, the
theatre closes its doors.

These four assets have unique characteristics. They do not bear a label
with the market price because they elude traditional accounting. Their
value is realized when they are intertwined in complex relationships, such
as those that might be forged with manufacturing. Unification between
manufacturing expertise and the aesthetic demands of opera can lead to
lyric opera productions enhanced by the application of complex digital

DOI: 10.1057/9781137492470.0019

technologies that connect people, processes, data and things. Equally, those in the audience could enrich their vision with augmented reality and communicate their emotions in real time.

There is a software program, MySmark, designed by an Italian, which creates an 'emotional tagging'. As such, customized profiles of members of the audience could allow theatres to investigate the personalities and subjectivities of those in audience, assess their feelings and understand their priorities. In offering this 'emotional footprint' measurement, theatres enter the world of intuition beyond the border of what the audience wants and journey into the territory of its hidden needs, to run across the grasslands of wilful ignorance.

Emotional paths: from Facebook 'Like' to MySmark emotional footprint 'Smark'

MySmark—the name is an abbreviation of 'Make your Smart mark'—is a digital innovation developed by an Irish start-up company at the junction of psychology–marketing–computer science and intended to provide web users with the ability to make richer and more powerful interactions online. If emoticons are more than 30 years old, well before the globalized Internet connections, users are still vague about when they express what they think and feel. A collection of many of these expressions, whether from an individual or a community, would be able to create an emotional path.

Many scientific and marketing teams have been working over the past two decades on multiple experiments to advance the capacity of understanding their users, with the aim of becoming smarter in capturing high-quality data and interpreting them meaningfully.

The era of mass-market connectivity and personal/business digitalization has increased the quantity of data generated by content providers as well as by end-users by an order of magnitude.

This critical mass of big data has caused new technologies to emerge in this domain—such as sentiment analysis and social listening. Both market(ing) niches investigate how to categorize and interpret user-generated content better, especially across social media and social networks, aiming to generate meaningful insights. Expert teams in these fields have reached a very high level of accuracy, albeit often expressed as generic outcomes such as

DOI: 10.1057/9781137492470.0019

'positive–negative–neutral', and the innovation opportunity seems to remain merely incremental.

Typically it is social networks that derive major advantages from the intersection of these technologies and big data. These networks can readily be at the forefront with regard to technology, choosing to develop it internally or search the market for best options. They can thus combine these analytical skills with a super-community of active users—as in the case of Facebook—keen on publishing content 24/7. Moreover they make use of so-called social plug-ins, a set of simple but straightforward call to actions that users can activate with a click: the Facebook 'Like' is probably the most famous, used—and misused—of them. It started as a powerful tool to create a positive feed (i.e., originally whatever a user liked was automatically published on their behalf), but it has since been sold to marketers and brand owners globally as the holy-grail of their online reputation and, ultimately, it has became somewhat ubiquitous as a type of web commodity.

But the most important thing proved by 'the Facebook Like' has been the widespread interaction brought to private, public and business material by the very large user base of the social network. In other words, it has been able to show user's engagement, preferences and recommendations, all with one click. It has given the social network giant a very powerful marketing tool for boosting its apparently exponential international growth.

What comes next seems to be an extension of the benefits of an audience's understanding to all business communities, from SMEs to corporates, from theatres to schools and bloggers. Targeted audience understanding consists of tangible and intangible components, thinking and feeling, objective and subjective insights. This is the main purpose of MySmark: providing its users with the ability to generate instant-rich-usable data on the customer experience. The alias of the social plug-in here is 'Smark–smart mark', a multidimensional bit left by the user during the journey when using or consuming content, a product or a service.

The corpus of 'Smark' contains the emotional footprint of the user, their personality, subjectivity and emotional profile.

Thus the emotional path during a concert, in an opera house, could become a set of marks concerning personality, colours and feelings, like a rose of emotions: Much more than a simple *like*.

DOI: 10.1057/9781137492470.0019

The 'Rose of Emotions' widget of MySmark

For the design and physical production of an opera, theatres could make use of 3-D printers. These and other new manufacturing technologies would offer experienced craftsmen, who create costumes, scenery and lighting for operas, opportunities to take their work further. Strong interaction with the world of manufacturing would enlarge the community of donors and investors in crowdfunding platforms. The eccentric profiles of 'nerds', who show a marked predisposition for science and technology, and 'geeks', who develop and enhance digital technologies using passion and experience, could complement the classic image of the opera connoisseur. Money of the many could give financial oxygen to theatres and shape an international community of opera lovers.

China, which has built its economic miracle on manufacturing, is in a quest for Grand Unification. The *National Grand Theatre of China* in Beijing, a giant egg-shaped building, can be likened to an Easter egg that contains the surprise of the design of dozens of new theatres, a melting pot of western-style and traditional Chinese opera. By investing in multimedia technologies, the Chinese aim to enhance their opera productions and performances.

The multiplier effect of opera houses

The opera, therefore, is becoming a non-elitist art form, with a very promising future. Current estimates by Bocconi, a private university in Milan, suggest that one Euro invested in *La Scala Opera House* generates two Euros in its supply chains and related industries. In the case of the *Teatro Comunale di Bologna*, a survey by Deloitte showed that for each Euro the *Comunale* receives in grants the community of Bologna

DOI: 10.1057/9781137492470.0019

gains benefits estimated to be around ten Euros, and each Euro in the budget of the *Comunale* generates 1.43 Euros of GDP in Italy, with significant beneficial consequences with regard to entrepreneurship and employment.

The long chain of activities required to stage an opera is full of opportunities to be exploited by innovative start-ups that combine technology with the intangible values unlocked by creative ignorance. That is why it would be necessary to promote and facilitate the role of the artist–entrepreneur and the technology-based artist who first explores the frontier that separates humans from machines and then creates interactions between entertainment and manufacturing. Returning again to the past, Beethoven was an artist entrepreneur; and in our lifetime there are technology-artists like Heather Knight, who works on theatrical robot performances. Here is a task for music conservatories and music academies which, as already happens in conservatories in the United States, should launch entrepreneurship courses for their students. As exemplified by the French programme *Dix mois d'école et d'Opera*, opera traces educational paths thanks to its connections with history, philosophy, literature, graphic arts, music, drama, and dance. Crossing cultural and national borders, opera houses multiply the economic impact of their performances. Here is a possibility: Italian opera programmes could revamp the *Grand Tour* of Italy in the light of the most acclaimed Italian composers.

Italy is home to the opera. Without passion and commitment to the Grand Unification, the country would be condemned to climb down the ladder of the world rankings for the number of opera performances per capita, as these data show.

Opera performances (annual average, 2013)	
Source: Corriere dell Sera, 04 October 2014	
In Italy	72.9
Staatsoper, Vienna	223
Metropolitan Opera House, New York	209
Royal Opera House, London	154

Today, placed on the twentieth step, opera houses in Italy live the paradox of poverty amid abundance, and the doctors at the bedside of the assumed moribund do not recognize their intangible assets. However, opera in Italy has the potential to become a benchmark and crossroads of creative talents once more. It should be remembered that each

DOI: 10.1057/9781137492470.0019

performance is a thread that unites the 'opera cities'; and because the markets are conversation, manufacturing would benefit greatly from teasing out the threads of the Grand Unification. 'I don't understand why people are afraid of new ideas, I am afraid of the old ones', said composer John Cage. If their intention is to eliminate the fear of novelty, opera houses might start by seeking the Grand Unification.

DOI: 10.1057/9781137492470.0019

12

Path Creators of the Third Millennium University

Formica, Piero. *The Role of Creative Ignorance: Portraits of Path Finders and Path Creators.* New York: Palgrave Macmillan, 2015. DOI: 10.1057/9781137492470.0020.

▶

DOI: 10.1057/9781137492470.0020

DOI: 10.1057/9781137492470.0020

To discover something really new, path creators of tertiary education in the twenty-first century decline to take into consideration aspects of the history of the university of the second millennium. The path creators, among whom entrepreneurial scientists and scientific entrepreneurs such as Jim Clark of Silicon Graphics and Netscape and Herbert Boyer of Genentech have been prominent since the beginning of this century, travel across the trans-disciplinary fields of education and research, leaving behind a silo-based educational mentality in order to give birth to ideas that can be shared across disciplines. In doing so, they establish privileged links and relations between teachers and students: a type of direct contract that does not have the public sector—the Ministry of Education, which sets criteria and procedures that educational institutions are obliged to follow—as an intermediary.

Within the framework of a rich and constant international flow of educators and learners, the modus operandi of the University of the Third Millennium is to promote and encourage the movement of ideas from one place to another; from natural sciences to the humanities and vice versa; from knowledge to creative ignorance; from research to entrepreneurship.

'Engaging with nothing, the unknown, the incomprehensible, and the unsaid.'

(University of Ignorance: Laetus in praesens)

An inconvenient oxymoron

Scientists and researchers proceed along two paths: one leads to publications in prestigious international journals and the other is aimed at enhancing entrepreneurially a scientific discovery, with its author giving birth to a new venture. However, some feel that combining science and business is something of an oxymoron. For many academics, researchers and entrepreneurs, science and business are, reciprocally, acute and obtuse. However, the entrepreneurial democracy that is emerging in the twenty-first century flourishes in the union between science and business. To stay out of it would be to condemn future generations to a long journey through the desert.

DOI: 10.1057/9781137492470.0020

Between the night and the dawn of scientific entrepreneurship

It is not sufficient to practise creative ignorance in curiosity-driven science laboratories. Creative ignorance in connection with blue-sky research must be introduced into university classrooms and into businesses. Stuart Firestein, director of the Department of Biological Sciences at Columbia University, teaches a course dedicated to creative ignorance and, answering questions raised by Casey Schwartz—a reporter with *The Daily Beast* (Schwartz, 2012)—Firestein draws attention to the fact that he

> Came to the realization at some point several years ago that these kids [his students] must actually think we know all there is to know about neuroscience. And *that's* the difference. That's not what we think in the lab. What we think in the lab is, we don't know bupkis. So I thought well, we should be talking about what we don't know, not what we know.

The joy of this process highlights the extent of what Firestein calls 'negative capability', 'the ability to remain in mysteries and unknowns without any irritable reaching or grasping'.

The pursuit of ignorance makes start-up founders curious and eager to see what happens once they escape from their maps of knowledge in order to occupy the space of not knowing. Above all, creative ignorance endorsed by path creators in different scientific fields ignites the flame of science-led entrepreneurship. In their absence, the timepiece of scientific entrepreneurship would indicate that it is late at night. As Anatole France wrote in his novel *Le Lys Rouge* (1894), 'What we see at night are the unhappy relics that we neglected while awake'. Indeed, many people fail to understand the high social and economic value of items in the treasure chest of 'Science and Entrepreneurship'. In classrooms and in university laboratories, while Marie Curie's genius and passion for scientific discovery are rightly admired, her entrepreneurial verve—the ability to raise money from private sources, to add to finance received from the public purse—remains largely neglected.

The taxing convergence of the 'two cultures'

More than 50 years have passed since the publication of *The Two Cultures and a Second Look* (Snow, 1963). The author, scientist C. P. Snow,

DOI: 10.1057/9781137492470.0020

denounced the lack of communication between the worlds of science and literature and the arts as being one of the most insidious evils of the twentieth century. In the group of path creators seeking to make connections between the two worlds, there are founders of start-ups who find themselves at the intersection between the two cultures.

Generally accepted beliefs, opinions and judgments are the unintended consequence of the production of knowledge: knowledge divided into a series of communities of closed doors (the 'silo syndrome') causes further reservations. Both are inhibitors of entrepreneurial processes that weave and merge disparate knowledge. From the United States to India and China, in the entrepreneurial universities of the Third Millennium, professors and students depart from silo mindsets.

This academic entrepreneurship *cum cursus honorum* is an alien concept to the Second Millennium universities. Organized by academic disciplines according to the canons of the nineteenth century, the universities of the second millennium are badly equipped for coaching researchers and students along the route that extends from knowledge gained in classrooms and laboratories to entrepreneurial action in the immense space of convergence of different scientific disciplines and humanities.

In Europe, and in particular in the Mediterranean countries, the debate continues about the morality of working with the worlds of industry and commerce, engaging on the one side supporters of the university that works closely with the industry and, on the other, those who support the university driven by the pure passion of discovery and devoid of such dangerous liaisons with businesses. As a result, universities do not acknowledge the transition area between the two species of neighbouring communities: academia, which generates knowledge, and industry which transforms knowledge into innovative ventures. There is much to be gained by these two communities, academia and industry, from the variety of species in the ecosystem, and from cooperation and the creative tension between them.

University reforms, of which so much is claimed and about which so much is praised, are not achieving this goal. When the playing field changes, redefinition of the university gains the upper hand over reform. This is proven by the rise and instant success of the Singularity University, Unreasonable Institute and Minerva Schools. Established by path creators in education, these three institutions have set in motion entrepreneurial processes between science and the humanities. The more

DOI: 10.1057/9781137492470.0020

the academic *cum cursus honorum* entrepreneurship front advances, the more universities entrenched in strongholds of the two cultures cannot maintain their position and are obliged to retreat.

The reimagined university

According to the predominant school of thought and what is considered the best current practice approach of research universities—a school that has among its most distinguished personalities Sir Leszek Krzysztof Borysiewicz, currently the 345th Vice-Chancellor of the University of Cambridge—the creation of a path that leads to the birth and development of science-based start-ups is a secondary effect, intended or accidental, of the discoveries and inventions derived from blue-sky research. Integration of the two paths and promptness in passing from one to another is a perspective recently evident in the new universities that are springing up in the third millennium.

The 'University of Ignorance' is the background of the reimagined university. Why that background? Part of the response lies in 'Identity, Possessive World-making and their Transformation Dynamics, 2012' (see Kairos@Laetus in-Praesens.org) where it is written that

> In a period when the quality of knowledge—so widely, arrogantly and uncritically hyped—would seem to be inadequate to the challenges of a global knowledge-based civilization, a more assiduous engagement with the ignorance implicit in such failure merits consideration. A more appropriate engagement with ignorance—and the process of ignoring—could prove to be a fruitful complement to the pattern of obsession with knowledge and its acquisition.

Another widely accepted aspect of the response lies in the wealth of human resources and diversity of talents, and their ability to interact and fertilize the changing educational terrain, all of which makes a country prosperous—a lesson that the United States has learned and long since made its own.

In Boston, Massachusetts, in 1861, and despite the presence of a cumbersome *prima donna* that was and still is Harvard University, founded in 1636, the Massachusetts Institute of Technology (MIT) was established. MIT was not intimidated by the presence of Harvard, nor did the *prima donna* manage to overwhelm the newcomer. Since that time they coexist in an environment of competition and cooperation. In

DOI: 10.1057/9781137492470.0020

the 1980s and 1990s, American corporations give birth to start-ups in education—the so-called corporate universities. The aim was to derive utility from academic research and the resultant industrial training. In the current century, scientists, inventors and innovative entrepreneurs are joining forces to reinvent the university. Their goal is to motivate the 'knowers' to run across the immense and unexplored prairies of creative ignorance so that they can enter into unexpected entrepreneurial territories.

The Californian Singularity University, founded in 2009, has quickly put on the clothes of a *prima donna*, a leading player in several countries. Backed by the founders of companies such as Google and Cisco, the Singularity University harnesses the power of technology, growing exponentially, to solve the great challenges of humanity. As stated in its vision, it is in the forefront of 'A global network of like-minded entrepreneurs, technologists and young leaders to participate in crafting a road map to guide the evolution of disruptive technologies'. Young people who keep an open mind are active at the crossroads between science and entrepreneurship: they are given guidance to set up innovative ventures in collaboration with scientists, academics and business leaders.

The ship of the Unreasonable Institute, based in Boulder, Colorado, carries nascent entrepreneurs from one port to the next of the most entrepreneurial cities on the planet. As the ship crosses the ocean routes, participants on board sail metaphorically with academics, serial entrepreneurs and innovators along the strait which links the sea of knowledge and the sea of creative ignorance.

The year 2014 is the year of the birth of Minerva Schools, the result of the coupling between digital technology and the innovation of the academic model of education. Like the Roman goddess who presided over not only war but also intellectual activities, Minerva Schools challenge traditional universities, reinterpreting their role in an imaginative way.

Minerva Schools recruit the *clerici vagantes* (wandering students) of the twenty-first century. They live anywhere in the world and enrol in online courses. In San Francisco, during the first six months of their course, Minerva students together develop a sense of community spirit. Thereafter, for the next six semesters they will travel to six cities in six different countries: Istanbul, Rio de Janeiro, London, Berlin, Mumbai, Hong Kong and Shanghai. Thus Minerva Schools are preparing to forge a student community without borders, engaged in the learning of advanced statistics, behavioural economics, computer science, and

DOI: 10.1057/9781137492470.0020

more: a community of young people, ambitious and talented enough to succeed in a traditional elite school. With tuition fees in the order of US$10,000 per year and with teachers recruited from the prestigious and elite private universities in the United States, Minerva Schools also differ from most American colleges with regard to its business model.

Minerva Schools: The 'Reimagined University'

'San Francisco-based technology executive Ben Nelson has set out to kick Harvard off its pedestal. Nelson, 38, is launching a new college via the Minerva Project, which he refers to as the "reimagined university", which is an alternative experience for those driven enough and talented enough to succeed in a traditional, elite school.

Students can live anywhere in the world and enrol in courses online. Now his vision is becoming a reality via the Minerva Schools at KGI, a joint venture between Nelson's Minerva Project and the Keck Graduate Institute, a member of the Claremont University Consortium in Southern California.'

Source: 'This entrepreneur is trying to create a "Perfect University" to displace Harvard and Yale—Minerva Chief Ben Nelson, *Innovation Daily*, January 8, 2014.

From the 'Nations' of wandering students to the 'International League' of wandering entrepreneurial Millennials

The Second Millennium University whose cradle is in Bologna—the University of Bologna—arose from the creative ignorance of wandering students grouped by 'nations'. Today, as globalization gives impetus to the Grand Tour of the World, the generation-wide movement of students relies on creative ignorance to open up borderless channels of communication. The interweaving of knowledge and its organization into social and business networks that transcend geographical, cultural, religious and ethnic barriers helps students to learn and exploit the differences, creating benefits for all.

Young, nascent entrepreneurs with tertiary-level education and highly varied cultural, social and geographic roots meet together on digital

DOI: 10.1057/9781137492470.0020

platforms, physically and virtually along new routes, in their race toward innovative entrepreneurship, and create borderless global start-ups with founders from different countries and continents. Thus does the 'Millennial Entrepreneurs' International League' emerge—a collection of wandering entrepreneurs in a unified culture, the latter-day version in the knowledge age of the Workers' International League of the industrial age of two separate cultures. While digital technologies caused jobs to be lost, creative ignorance promotes and encourages cultural and entrepreneurial convergence, thus paving the way for path creators in the twenty-first century.

The 'Bologna syndrome': how Bologna 'the learned' is being engulfed by relational capitalism

Bologna 'the learned'—so-called because the city is home to the University of Bologna, the *Alma Mater Studiorum*, founded in 1088—is a cornerstone of relational capitalism in Italy. The title 'the learned' acknowledges that the city has gained deep and thorough knowledge thanks to its university, considered to be the oldest in the Western world. 'Relational capitalism' is a form of capitalism in which the members of a community set up various and varied relationships and affiliations in order to establish a network characterized by exclusivity and a lack of receptiveness to external members—unless they are specifically invited to participate. In short, relational capitalism is an oligarchy answerable only to, and feeding only, itself.

How did it happen that the 'learned' city has compromised the meritocratic principle of what you know—the 'know-how'—so much and, conversely, has rewarded beyond all measure the relational principle of who you know—the 'know-who'? In their attempts to answer this question, experts in the field of relational capital have addressed many causes that supposedly lie at the root of the problem. However, it seems there is one cause that has been ignored and abandoned. In its pioneering times, the University of Bologna was a free and secular organization with regard to its students who, as the 'learners', ruled the *Alma Mater Studiorum*.

It was 'learned ignorance' that gave rise to new knowledge: so it was that, in such a cultural environment, Bologna became famous in Italy and beyond as 'the learned'. When the governance of the university subsequently passed into the hands of the university professors, their power as keepers of knowledge enabled them to take control of the

DOI: 10.1057/9781137492470.0020

process of student learning in Bologna—and indeed, in similar fashion, in many other universities. Through the exercise of that power, repositories of knowledge became established, to become the foundations on which would stand the edifices of professional bodies (with the professors themselves as influential members), business and trade associations, trade unions and public sector bureaucracy, all gaining ever-increasing importance in contemporary Italian society. Having the right contacts, the 'know-who' thus becomes a crucial factor when society is in the grip of bodies such as these, acting in collusion and exercising their power to authorize, prohibit, limit and regulate the many aspects of everyday life.

Bologna, 'the learned', is a vivid testimony of the unintended consequences that knowledge which rejects the learned ignorance can produce. The pursuit of knowledge leading to relationship capitalism is the cause of what we call the 'Bologna syndrome', a psychological phenomenon that sees the learners as hostages of the knowers, the captors. Forced or, in a burst of sympathy, wishing to keep relations with the captors to avoid the ostracism and, conversely, to get the benefits conferred by cronyism, the learners replace the baggage of the 'know-how' with that of 'know-who'. That syndrome could be defeated by returning to the original principle of Bologna 'the learned'.

This, however, could only be done through the creation of another university entirely new to that founded in 1088. The Universities of the Third Millennium are already emerging from the partnerships between scientists and entrepreneurs, with the pro-active participation of students. Starting from Bologna, the landscape of tertiary education in Italy would be thoroughly re-thought out in the light of the new universities. But how to do it, without severing by sword the Gordian knot of relationship capital?—whose anticompetitive and non-meritocratic principles and practices do not permit another tertiary-level institution to arise in competition, adjacent to the ancient and glorious university and beyond the fence of collusive relations. Will the 'first-generation' entrepreneurs (i.e., entrepreneurs who have created a company for the first time), including the majority of the founders of innovative start-ups, be the ones to take up the sharp sword? They might do so if their companies will enter into territories inaccessible to relationship capitalism unable to learn the lasting and irrevocable lessons of the great upheavals in science and technology which have significant social and economic consequences.

DOI: 10.1057/9781137492470.0020

Epilogue: Traveling to the No Comfort Zone and Reaching the Flow Zone

Formica, Piero. *The Role of Creative Ignorance: Portraits of Path Finders and Path Creators*. New York: Palgrave Macmillan, 2015. DOI: 10.1057/9781137492470.0020.

▶

DOI: 10.1057/9781137492470.0020

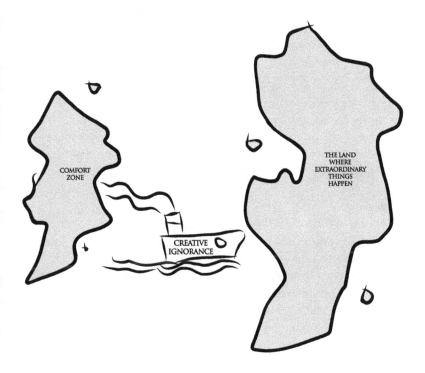

DOI: 10.1057/9781137492470.0020

The sense of reality makes us see the world as it is. Ideas not consistent with reality seem far-fetched and irrelevant, if not despised, until the 'ugly duckling' fails to turn into a 'swan'. Knowledge is imbued with a sense of reality, and, thus, its followers—the path finders. In contrast, creative ignorance relies on the sense of possibility. As Robert Musil (1995) wrote,

> *Whoever [as we would say here, the creative ignorant seen as a 'man without qualities'] has it does not say, for instance: Here this or that has happened, will happen, must happen; but he invents: Here this or that might, could, or ought to happen. If he is told that something is the way it is, he will think: Well, it could probably just as well be otherwise. So the sense of possibility could be defined outright as the ability to conceive of everything there might be just as well, and to attach no more importance to what is than to what is not.*

Knowledge generates freedom and slavery concurrently. With their extensive knowledge, the path finders are free to take the risk of a new journey of exploration. However, that knowledge is a heavy burden that makes them slaves of the awareness of their own limits. The future is seen through the lens of risk and thus the path finders discard projects that knowledge suggests will be unrealistic, too ambitious, not within their reach.

In contrast, having abandoned the baggage of knowledge largely as a result of selective memory, the precursor of ignorance, the lodestar of intuition tells the path creators where to look next and creative ignorance gives them the freedom to embark upon projects—perhaps unsteadily and uncertainly at first—which, even if in a disorganized and unsystematic manner, erode the path finders' achievements.

However, no matter how different they are from each other, the two species of path finders and path creators, with interactions between them leading to reciprocal discussions, are an important part of the narrative that unfolds between the clarity of knowledge, which makes path finders run risks; and the doubt of creative ignorance, which enables path creators to become accustomed to the world of uncertainty.

> *So you have to force yourself out of a comfort zone and really try to figure out what are the key ingredients, the key skill sets, the key perspectives that are necessary, and then figure out a way to attract the very best people to fill those particular roles. (Steve Case, 2008)*

DOI: 10.1057/9781137492470.0020

Thoroughbreds and hopeful monsters

It is important to be relaxed, to get into shape for global competition in incremental innovation. For this reason path finders, rather like thoroughbreds horses, live in a comfort zone, accommodated on the bed of the knowledge accumulated throughout the course of their scientific, technological and business heritage.

In contrast, path creators begin a journey to the 'no comfort' zone and then reach the 'flow zone', as Ilkka Kakko (2014) has named it, a new and entirely different comfort zone. Here, more than anywhere else, path creators feel fully realized, being able to express their creativity where knowledge and creative ignorance are forcefully intermingled, as if the strong currents of two fast-flowing rivers. From this perspective, path creators are hopeful monsters, genuine mutants essential for creating from nothing pathways that lead entrepreneurship beyond the boundaries drawn on the knowledge maps. Path creators show a strong propensity to detach from statistically significant events, from what is commonly known as the 'world of reality', to become familiar with the 'world of the improbable', with facts and events seemingly insignificant today but, tomorrow, so important as to induce a profound change in the current state of affairs. Their ideas and intentions come to be seen as vague, bizarre and difficult to implement. The motivation that accompanies creativity backed by conviction makes them workable.

How path finders and path creators deploy knowledge and creative ignorance

	Knowledge	Creative ignorance
Path finders ('Thoroughbreds')	Knowing where to go in the 'world of reality' guided by tenets and dogmas of knowledge	Knowledge dogmas ignore or remain firmly opposed to counter-intuitive insights
Path creators ('Hopeful Monsters')	Not knowing where to go	Creative ignorance triggers a generative process of familiarizing path creators with the 'world of the improbable'

The genesis of a new era has the imprint of path creators. They are happy to be part of a world in which the lives of many will be changed, by, for example, the fight against Alzheimer's and dementia, and prevention

DOI: 10.1057/9781137492470.0020

of diabetes, by development of methods to halt their progress; solar energy generated even when there is no sunlight; electric aircraft; objects permanently connected to each other through the Internet of Things; and many more technological innovations. With novel ideas, path creators go to the scene of action, doing something disruptive to the legacy accumulated by the rulers of a bygone era. For them the opportunity to take action based on intuition and not discouraged by uncertainty, to trace entirely new paths, is much more rewarding and fulfilling than relying on extant knowledge to make existing markets more efficient and competitive. Unrestricted by the corset of knowledge and not obliged to deliver 'results' in the short term, path creators can navigate at length the waters of 'not knowing', to experience the delights of surprise at what happens, until truly creative results emerge.

The new era of consumer empowerment, arriving as a result of mobile and cloud computing technologies, has brutally exposed the dysfunctional culture of companies currently dominating the computer industry, in which most incumbents have yet to show signs of adapting to the 'post-your industry' world.

In the computer industry, as in others, incumbents are locked-in to the dominance of their industry prior to the transition to this new era. The lock-in syndrome hurts them twice. First, the adherence to 'business as usual' is a gravitational force strong enough to restrain them within the visible horizon. They resign themselves to the role of path finders within the range of investigation circumscribed by that horizon. Second, the syndrome causes losses of instinct, intuition and insight. These are the attributes plentiful among the outsiders—those fresh talents who, by creating new and more productive paths, go beyond the visible horizon.

In the 'Path Creation' theatre, it is by means of familiar characters like the iPod, the iPhone, the iPad and Kindle that the audience feels involved. However, popularity has features similar to those of the woman in Verdi's *Rigoletto*: she 'is flighty like a feather in the wind, she changes her voice—and her mind'. Creative ignorance is a strong wind that makes the feather fly away. When the doors of the theatre open to welcome novel and deliberately ignorant creators, other characters enter the scene and replace the old ones who thus lose the luxury of pointing the way ahead and are happy to encourage the audience to follow them. This is the situation in which Steve Jobs found himself with regard to Google, as did the new CEO of Blackberry. Apple now seems more interested in protecting its accumulated knowledge than in reviving the creative

DOI: 10.1057/9781137492470.0020

ignorance of the past, whereas former Blackberry consumers want what the company currently does not have.

At the border of the comfort zone

Within the boundaries of the comfort zone demarcated in their knowledge maps, path finders go in search of as-yet untrodden paths. Travelling along new routes, they do 'more of the same' for more people. The goal is to reform the current state of affairs through incremental innovation, in the name of continuity.

The mission of path creators follows a completely different path. Disobedient to the holders of knowledge, abandoning the rules of sacred textbooks, they create—from start to finish—paths that lead them to discontinuous changes. In doing so, path creators redefine the current state of affairs. 'More of the same' gives away to 'something completely different from the same'.

In his play *No Man's Land* Nobel Prize-winning English playwright Harold Pinter wrote: 'You are in no man's land. Which never moves, which never changes, which never grows older, but remains forever, icy and silent.'

But it is not always so. Tensions and frequent guerrilla warfare can be observed in the no man's land that separates path finders from path creators. Sometimes it happens that war and peace coexist in a unique and inextricable manner, as in the case—now legendary—of the rivalry between Sony and Philips for the development of laser technology that has revolutionized, among other things, listening to recorded music.

Sony and Philips: sometimes allies, at other times rivals in redefining listening to recorded music

The case of the 'war and peace' between Sony and Philips in the 1980s has become well-known in the music industry. Comprehensive accounts were published in the Wall Street Journal on 05 March 1986 and 23 January 1987, in two articles signed, respectively, by E. S. Browning and Mark Nelson. The following is a version of the case adapted to the needs of our present narrative.

At the beginning of the 1970s a Sony researcher, studying the relevant literature, learned about experiments that Philips was carrying

DOI: 10.1057/9781137492470.0020

out on the use of laser technology for listening to recorded music. The reaction of the Japanese corporation was immediate, but the proverbial patience of the people of that nation was exhausted by the presence of a barrier they were unable to overcome: finding a type of disc better suited to the new technology. Sony's engineers failed to conceive the mini-disc—although they had already made other 'mini' products such as televisions—and remained attached to the concept of a disk the size of a long-playing vinyl record that would be able to record, using the new technology, 15 hours of music. Unfortunately, this was not a product that could be offered to the market at a reasonable price.

In 1977, Sony decided to suspend its experiments. Some ten months or so later, something unexpected occurred. Some managers employed by Philips went to Tokyo to demonstrate their laser-based audio CD (originally, some 115 mm in diameter) to Sony and Matsushita. The Dutch offered a cooperation agreement to jointly promote the new technology; and Sony, unlike Matsushita, accepted. The agreement provided for simultaneous entry into the market, starting in October 1982.

Building on the innovative potential arising from the agreement with Philips, Sony committed itself to a mini version of the optical drive, gambling—with an aggressive marketing policy—on the importance of the price difference of about US$500 between the regular (US$800) and mini (US$300) versions. Sony flooded the US market, putting strong pressure on Philips and, as a result, the Dutch company, in the eyes of consumers, lost its image as the pioneer of new technology. Sony assumed that status and, among other things, compelled established companies such as Zenith and RCA to abandon the market. Furthermore, Philips was obliged to ask for protective tariffs in the EU Common Market.

The interaction between Sony and Philips was crucial for crossing this particular stretch of no man's land. Sony was victorious because the Japanese led the way with applications of a new technology capable of generating a mass market. In contrast, the Dutch opponent was seemingly too enamoured of success achieved in research laboratories, overestimating the technology and underestimating the most promising applications on the market. Rather than freezing the competition, the interaction between Sony and Philips triggered

DOI: 10.1057/9781137492470.0020

a competitive process which resolved in favour of the company that chose to go on the attack by differentiating its product.

Along that strip of land path finders and path creators meet and collide. Some lessons can be learned from the encounter between Sony and Philips, lessons that have found resonance in the work of Xuan Pham in *Five Principles of Path Creation* (2006–2007). Business leaders are not self-sufficient and, therefore, surround themselves with people—among them, those possessed of the scientific and technical skills necessary to design and implement the infrastructure of the new paths—who are committed to contributing to their leader's work. Those entrepreneurs who manage to cross no man's land are ready to pave flexible paths—that is, paths that provide deviations and ways of exiting. Although rivals, the protagonists share information spaces and are very selective about the volume, quality and types of data to be transmitted to rivals.

At stake is primacy in the entrepreneurial economy. According to the Global Entrepreneurship Barometer of the Mason Centre for Entrepreneurship and Public Policy, the world is currently (2014) operating at 25% of entrepreneurial capacity. This is encouraging in the presence of the deep economic crisis of recent years and because the new entrepreneurs create a virtuous circle of confidence in the markets necessary for renewed economic growth and increased levels of employment. If the confrontation between the two sides takes the form of 'co-opetition'—a competitive and at the same time cooperative approach to conflict—we will see the emergence of a greater number of higher quality dense and connected networks of relationships to facilitate the birth of international communities including policy makers, old and new entrepreneurs, serial entrepreneurs, mentors, investors, universities, and governments. All of these will together be committed to increasing the number of successful, high-flying start-ups, thereby generating income and employment opportunities, in contrast to those that stagnate in the lower layers of entrepreneurship.

Path finders are willing first to tread and then to conquer no man's land in order to plough the field of incremental innovations expected to enhance the productivity and competitiveness of existing businesses, in the hope and expectation of positive side-effects on corporate management and employment.

DOI: 10.1057/9781137492470.0020

In contrast, path creators aim to take over no man's land in order to enrich the biological diversity of the species that populate that particular economic landscape. Their mission is entrepreneurial rather than managerial. With innovative start-ups that are more open and flexible in the face of increasingly dynamic consumers whose expectations are changing rapidly, entrepreneurial species and the business population tend to increase in numbers and try to find a way through a world of uncertainty.

The effect of therapy applied by path creators is clearly visible in the Netherlands, today perhaps the most entrepreneurial country in the European Union with 7.2% (4.9% in 2001) of the population between the ages of 18 and 64 having founded a start-up or being in the process of doing so. Increasing numbers of citizens of the Netherlands believe they have the right skills to do business and the increasing number of entrepreneurs creates a domino effect: every new-born entrepreneur raises the propensity to select entrepreneurship as a good career choice.

In Europe, advances and setbacks on the two sides of the game have coincided with a now lengthy period of unemployment and inactivity, while on a global scale the level of entrepreneurship is rising. The higher this level, the better the opportunities to plough new entrepreneurial fields, to open the doors of work to a world population of over seven billion. At the start of the second decade of this century, The Global Entrepreneurship Monitor has recorded 400 million entrepreneurs in 54 countries, with millions of new jobs expected in the years to come. In the United States, the Millennial generation—young people born between the dawn of the 1980s and the beginning of the new Millennium—will increasingly play an entrepreneurial role, or may already have launched a start-up. The Millennials discard the idea of 'safe' and wear hybrid clothing, made with entrepreneurial fabric and designed using temporary business experiences gained in the spirit of an entrepreneurial mind.

In the European Commission headquarters in Brussels, commissioners say they need to reinvent the entrepreneurial spirit in Europe and infuse it in the public sector. In fact, the propensity for entrepreneurship is in sharp decline. The preference for employment is more pronounced: 58% of Europeans want this, compared to 49% a few years ago. The Global Entrepreneurship Monitor reports that in the international competition for innovation whose outcome is entrepreneurship, from the advanced economies the United States lined up a group of new entrepreneurs

DOI: 10.1057/9781137492470.0020

equal to 7.8% of the population, compared to 4.2% for Germany; among the emerging nations Brazil and China lead the field with 14% and 17%, respectively. Hence the hope arises that an entrepreneurial European Union will be born (the 'EU' as an 'Entrepreneurship Union', as the UK says), a group of meritocratic 'Start-up Nations' greater than the sum of their parts.

In countries such as Italy where path creators, starting with those of the first generation, face high barriers raised by path finders of the second, third and fourth generations, the group of new entrepreneurs is small, a mere handful at 2.3% of the Italian population. The performance of the entrepreneurial economy in Italy is placed 27th among the 70+ countries as classified by the Global Entrepreneurship and Development Index. The Italian score is 60 percentage points below the highest mark and drops further for the two sub-indices of entrepreneurial aspiration and entrepreneurial activity. In short, the country is struggling to launch innovative start-ups in medium- and high-technology-based sectors, founded by well-educated new entrepreneurs. According to the Global Entrepreneurship Monitor 2012, Italy is at the bottom of the European league table for quality of innovative entrepreneurship, with a modest number of start-ups acting as employment accelerators because they are able to grow significantly. Nevertheless, data released by *Startup Italy*, the association of Italian start-ups, show that the potential for entrepreneurship is not negligible, with 300,000 aspiring entrepreneurs from 5.3 million individuals interested in investing in their business plan as well as in other people's projects.

In the other competition for primacy—the creation of new jobs—Europe is expected to compete by choosing its champions according to the criterion of size rather than age. However, OECD research has shown that it is not the size but the age of firms that makes the difference. It is young and innovative companies and particularly those experiencing or with the potential for high growth, operating across a wide range of industries, that create more jobs. In the United States the net creation of jobs is much higher among young firms than mature businesses. In the first group, four out of ten posts being filled are new jobs; in the second group, the number is less than one in ten (between 0.25 and 0.33, in fact). This performance has convinced the US administration that innovative start-ups are the champions in the race, nurtured with 50% of expenditure on R&D in contrast to a meagre 7% in Europe. However, the difference is not restricted to expenditure on R&D and innovation.

DOI: 10.1057/9781137492470.0020

Accelerated by four factors, the velocity of circulation of entrepreneurship is at least as important. These four factors are:

▸ Universities that move knowledge from its point of origin to the entrepreneurial channel;
▸ Existing companies that are incubators of innovative start-ups;
▸ Corporate downsizing, resulting in staff involved to be strongly motivated to consider entrepreneurship; and
▸ Fledgling entrepreneurial heroes whose achievements are a source of new opportunities to be exploited by aspiring and new entrepreneurs.

Are entrepreneurs, whether path finders or path creators, replacing employees in the New World of the Millennium? In the crowd of entrepreneurs that is developing during these first two decades of the new century, we can begin to discern, next to the path finders, first-generation path creators—innovators and leaders of innovative entrepreneurship who redefine the content and context of the economy. They will play a key role in the entrepreneurial democracy having the imprint of a biodiversity economy. Their attitudes and forcefulness in doing business produce the energy that is exploited by entrepreneurial ecosystems. Google and Intel are two of the many examples of ecosystems that raise and finance the game changers of the Millennium. Their passions and aspirations attract another crowd, that of crowdfunding—the many people who decide to invest small amounts in budding entrepreneurs, either for social solidarity or because they are driven by curiosity to test new products, or even by the possibility of deriving an economic benefit. It is dangerous today to argue that the entrepreneurs will be raised on the altars while the employees will end up biting the dust. The age in which we are immersed is not the sole preserve of entrepreneurs. At the border of the comfort zone, in the midst of tensions, guerrilla warfare and co-opetition, path finders and path creators who cultivate the field of entrepreneurship with a wide spectrum of economic views have both to contend not only with one another but also with employees.

Interact and negotiate

It is in no man's land that practices of interaction are being tested; and from this interaction inter-company agreements are negotiated. While

DOI: 10.1057/9781137492470.0020

the game for leadership needs to be played, negotiation produces a result that satisfies a large proportion, though by no means all, of the interested parties: that is, the fact that the market will expand in an orderly manner and thus be more promising for all. It was this that was the goal achieved by Philips through its strategic decision to negotiate with Sony. Although forced to navigate turbulent waters, the Dutch company managed to retain a leading position in a dynamic market.

Interactions and negotiations between path finders, between path creators and among the former with the latter are witness to the fact that innovation is the result of contamination and cross-fertilization between all parties involved. Interaction is a game that requires not only education but also creative ignorance. In the field of education there are the supporters of an educational system that focusses on science, technology, engineering and mathematics, scientific progress being considered the engine of innovative ideas. In the field of creative ignorance, the passions, desires and resourcefulness of the people are the vehicles of innovative ideas. As the Nobel Prize winner in economics Edmund Phelps (2014) suggests, it is the population's dynamism rather than the progress of science which, in the 1920s, enabled the United Kingdom and the United States to become the most innovative economies, followed in the course of the twentieth century by Germany and France.

If, as the polymath Benjamin Franklin said, ignorance is more expensive than education, it is nevertheless also true that being overconfident about one's own knowledge may prove to be even more detrimental. Being aware of this and behaving accordingly, innovation, then, is an infinite game between the path-finding entrepreneurs who, with an eye on the consequences of their actions, move along trajectories traced by their knowledge, and the path-creating entrepreneurs who, with an eye on what motivates them, embrace creative ignorance in order to enter uncharted territories.

DOI: 10.1057/9781137492470.0020

Selected Bibliography

Anderson, M. (1992), *Imposters in the Temple: American Intellectuals are Destroying Our Universities and Cheating Our Students of Their Future*, New York, NY: Simon and Schuster.

Bartholomew, J. G. (1911), *A Literary & Historical Atlas of America*, London, UK: J. M. Dent & Sons.

Blank, S. (2014), 'Steve Blank on the next 50 years of business innovation', *Innovation Daily*, 14 April.

Boschma, R. A. (2005), 'The role of proximity in interaction and performance: Conceptual and empirical challenges', *Regional Studies*, 39(1): 41–45.

Bryson, B. (2003), *A Short History of Nearly Everything*, London, UK: Black Swan Books, Chapter 27.

Case, S. (2008), Steve Case: Interview, *Academy of Achievement*, available at: www.achievement.org/autodoc/page/cas1int-4.

Castaneda, C. (1970), *The Teachings of Don Juan: A Yaqui Way of Knowledge*, London, UK: Penguin Books.

Cavafy, C. P. (2007), *The Collected Poems*, Oxford, UK: Oxford University Press.

Ceccarelli, G. and Grandi, A. (2011), 'A "Made in Italy" made by the English: How Marsala wine became a typical Italian product', paper presented at the conference '*Connections and Comparison – Third European Congress on World and Global History*, London, 14–17 April.

Christensen, C. (1997), *The Innovator's Dilemma: When New Technologies Cause Great Firms to Fail*, Cambridge, MA: Harvard Business Press.

DOI: 10.1057/9781137492470.0022

Curley, M. and Formica, P. (2013), *The Experimental Nature Of New Venture Creation: Capitalizing on Open Innovation 2.0*, Berlin, Germany: Springer.

de Finetti, B. (1990), *Theory of Probability*, 2 volumes, New York, NY: Wiley and Sons.

de Grasse Tyson, N. (2005), 'The perimeter of ignorance. A boundary where scientists face a choice: invoke a deity or continue the quest for knowledge', *Natural History Magazine*, November.

de Saint Exupéry, A. ([1943] 1995), *The Little Prince* (English-language Wordsworth Edition), Ware, UK: Wordsworth Editions.

Dewey, J. (1933), *Essays and How We Think*, Revised Editions, Boydston, J. A. ed, (1986, 2008), Carbondale, IL: Board of Trustees, Southern Illinois University.

Drucker, P. (1964), *Managing for Results*, New York, NY: Harper & Row.

Drucker, P. F. ([1985]1986), *Innovation and Entrepreneurship: Practice and Principles*, Perennial Library Edition, New York, NY: Harper & Row.

Eisenmann, T.R. (2013), 'Entrepreneurship: A working definition', *Harvard Business Review*, 10 January, available at: https://hbr.org/2013/01/what-is-entrepreneurship/.

Feyerabend, P. (1975), *Against Method*, London, UK: New Left Books.

Firestein, S. (2012), *Ignorance: How it Drives Science*, Oxford, UK: Oxford University Press.

Formica, P. (2013), *Stories of Innovation for the Millennial Generation: The Lynceus Long View*, Basingstoke, UK: Palgrave-Macmillan.

Franklin, C. (2003), *Why Innovation Fails: Hard Won Lessons from Business*, London, UK: Spiro Press.

Freakonomics (2009), *Survivor Bias on the Gridiron*, 17 September, available at: http://freakonomics.com/?s=Survivor+Bias+on+the+Gridiron&x=0&y=0

Frydman, R. and Golberg M. D. (2007), *Imperfect Knowledge Economics: Exchange Rates and Risk*, Princeton, NJ: Princeton University Press.

Fuchs, C. A. (2010), 'QBism, the Perimeter of Quantum Bayesianism', *Quantum Physics*, Ithaca, NY: Cornell University Library, 26 March.

Gelatt, H.B. (2008), 'In Search of Ignorance: Cultivating a Beginner's Mind', available at: http://www.gelattpartners.com/images/In_Search_of_Ignorance_by_H_B_Gelatt_070708.pdf (accessed 02 December 2014).

Geroski, P. (2003), *The Evolution of New Markets*, Oxford, UK: Oxford University Press.

DOI: 10.1057/9781137492470.0022

Goethe, J. W. (2010), *Italian Journey*, London, UK: The Folio Society.
Goldschmidt, R. (1940), *The Material Basis of Evolution*, New Haven, CT: Yale University Press.
Gruel, W. (2014), *Open Innovation und Individuelle Wissensabsporption*, Wiesbaden, Germany: Gabler.
Handy, C. (1989), *The Age of Unreason*, London, UK: Hutchinson.
Hazlitt, W. (1821), *Table Talk: Essays on Men and Manners*, London, UK: Henry Frowde.
Herbert, F. (1965), *Dune*, London, UK: Hodder.
Ito, J. (2014), Want to Innovate? Become a 'now-ist', *TED Talk Video*, available at: http://www.ted.com/talks/joi_ito_want_to_innovate_become_a_now_ist/transcript
Jones, C. P., ed., (2005), *Philostratus: The Life of Apollonius of Tyana*, Cambridge, MA: Harvard University Press.
Kakko, I. (2004), *Oasis Way and the Post-Normal Era – How Understanding Serendipity Will Lead You to Success*, St. Petersburg, Russia: BHV.
Kalbach, J. (2012), 'Clarifying innovation: Four zones of innovation', *Experiencing Information*, 3 June, available at: http://experiencinginformation.wordpress.com/2012/06/03/clarifying-innovation- four-zones-of-innovation/
Keynes, J. M. (1930), 'Economic Possibilities for our Grandchildren II', *The Nation and Athenaeum* 48(2), 11 October: pp. 36–37, later reprinted in *Essays in Persuasion*, London, UK: Macmillan (1933).
Leopardi, G. (2013), *Zibaldone: The Notebooks of Leopardi*, Caesar, M. and D'Intino, F., eds, London, UK: Penguin Classics.
Machado, A. (2003), *Border of a Dream: Selected Poems of Antonio Machado*, Barnstone, W., translator, Port Townsend, WA: Copper Canyon Press.
Manyika, J., Sinclair, J., Dobbs, R. et al. (2012), *Manufacturing the Future: The Next Era of Global Growth and Innovation*, available at: http://www.mckinsey.com/insights/manufacturing/the_future_of_manufacturing
Merton, R. K. and Barber, E. G. (1992), *The Travels and Adventures of Serendipity: A Study in Historical Semantics and the Sociology of Science*, Bologna, Italy: Società Editrice il Mulino.
Musil, R. (1995), *The Man Without Qualities*, New York, NY: Knopf.
Nocera, J (2013), 'How not to stay on top', *New York Times*, 30 August.
Petit, P. (2012), *Cheating the Impossible: Ideas and Recipes from a Rebellious High- Wire Artist*, New York, NY: TED Conferences, see also: www.ted.com.

DOI: 10.1057/9781137492470.0022

Pham, X. (2006–2007), 'Five principles of path creation', *Oeconomicus*, Volume VIII.

Phelps, E. S. (2013), 'Meno innovazione, più disuguaglianza', *Il Sole 24 Ore*, 24 July (excerpt from Phelps, E. S. (in press), 'Mass flourishing: How grassroots innovation created jobs, challenge and change').

Phelps, E. S. (2014a), 'Corporatism not capitalism is to blame for inequality', *Financial Times*, 25 July.

Phelps, E. S. (2014b), 'Teaching economic dynamism', *Project Syndicate*, 02 September.

Popper, K. (1957), *The Poverty of Historicism*, London, UK: Routledge.

Rescher, N. (2009), *Ignorance: On the Wider Implications of Deficient Knowledge*, Pittsburgh, PA: University of Pittsburgh Press.

Rogers, E. M. (2003), *Diffusion of Innovations*, 5th edition, New York, NY: Free Press.

Scandling, M. (1968), '2 of founders leave Fairchild, form own electronics firm', *Palo Alto Times*, 02 August.

Schwartz, C. (2012), 'Stuart Firestein, author of "Ignorance", says not knowing is the key to science', *The Daily Beast*, 22 April.

Scott, M. (2009), *From Democrats to Kings: The Brutal Dawn of a New World from the Downfall of Athens to the Rise of Alexander the Great*, London, UK: Icon Books.

Skidelsky, R. (1992), *John Maynard Keynes, II: The Economist as Saviour, 1920–1937*, London, UK: Macmillan.

Smith Purton, J. (1886), *M.T. Ciceronis Oration Pro Tito Annio Milone, with a Translation of Asconius: Introduction, Marginal Analysis and English Notes*, Cambridge, UK: Cambridge University Press.

Sobel, D. (1995), *Longitude: The True Story of a Lone Genius Who Solved the Greatest Scientific Problem of His Time*, New York, NY: Walker & Company.

Sobel, D. (2014), 'The lost pages of "Longitude" ', *Financial Times*, 15 February, 18.

Taleb, N. (2012) *Antifragile: Things That Gain from Disorder*, New York, NY: Penguin Random House.

Thomke, S.H. (2003) *Experimentation Matters: Unlocking the Potential of New Technologies*, Boston, MA: Harvard Business School Press.

Trevelyan, R. (1972), *Princes Under the Volcano. Two Hundred Years of a British Dynasty in Sicily*, London, UK: Orion, also Macmillan.

Tjan, A. K. (2010), 'The power of ignorance', *HBR Blog Network*, 09 August.

DOI: 10.1057/9781137492470.0022

Verdoux, P. (2009), 'Towards a theory of ignorance', *Philosophical Fallibilism Blog*, 20 March.

Westlake, S. (2013), 'The West desperately needs more madcap schemes like the Hyperloop', *Financial Times*, 17 August.

Williams, J. E. (1972), *Augustus*, New York, NY: Viking Press.

Wright, R. (2004), *A Short History of Progress*, New York, NY: Carroll and Grad Publishers.

Žižek, S. (2006), 'Philosophy, the "unknown knowns", and the public use of reason', *Topoi*, 25(1–2): 137–142.

Zweig, S. (2009), *The World of Yesterday*, London, UK: Pushkin Press.

DOI: 10.1057/9781137492470.0022

Index of Names

DOI: 10.1057/9781137492470.0023

DOI: 10.1057/9781137492470.0023

DOI: 10.1057/9781137492470.0023

GPSR Compliance
The European Union's (EU) General Product Safety Regulation (GPSR) is a set
of rules that requires consumer products to be safe and our obligations to
ensure this.

If you have any concerns about our products, you can contact us on

ProductSafety@springernature.com

In case Publisher is established outside the EU, the EU authorized
representative is:

Springer Nature Customer Service Center GmbH
Europaplatz 3
69115 Heidelberg, Germany